McDougal Littell

Biology

Ecology
Unit 5
Resource Book

McDougal Littell
A HOUGHTON MIFFLIN COMPANY
Evanston, Illinois • Boston • Dallas

ISBN-10 0-618-72532-6

ISBN 978-0-618-72532-8

4 5 6 7 8 9 - MJT - 10 09 08

Ecology:
Unit 5 Resource Book

CHAPTER 13 PRINCIPLES OF ECOLOGY

CHAPTER 14 INTERACTIONS IN ECOSYSTEMS

CHAPTER 16 HUMAN IMPACT ON ECOSYSTEMS

UNIT 5 ECOLOGY

Name Period Date

SECTION
13.1 | ECOLOGISTS STUDY RELATIONSHIPS
Study Guide

KEY CONCEPT
Ecology is the study of the relationships among organisms and their environment.

VOCABULARY	
ecology	ecosystem
community	biome

MAIN IDEA: Ecologists study environments at different levels of organization.
Write a description of each level of organization in the table. Also, provide an example for each level.

Level	Description	Example
1. organism		
2. population		
3. community		
4. ecosystem		
5. biome		

MAIN IDEA: Ecological research methods include observation, experimentation, and modeling.

6. What is observation?

7. What is the difference between direct and indirect surveys?

8. Complete the following table with a benefit and drawback of conducting an experiment in the laboratory compared with conducting an experiment in the field.

Experiment	Benefit	Drawback
Laboratory		
Field		

9. When might a scientist use a model as a research method?

Vocabulary Check

10. What is ecology?

11. Of the three terms, *biome*, *community*, and *ecosystem,* which term contains the other two?

Name **Period** **Date**

SECTION **13.1** | ECOLOGISTS STUDY RELATIONSHIPS

Power Notes

Levels of Organization

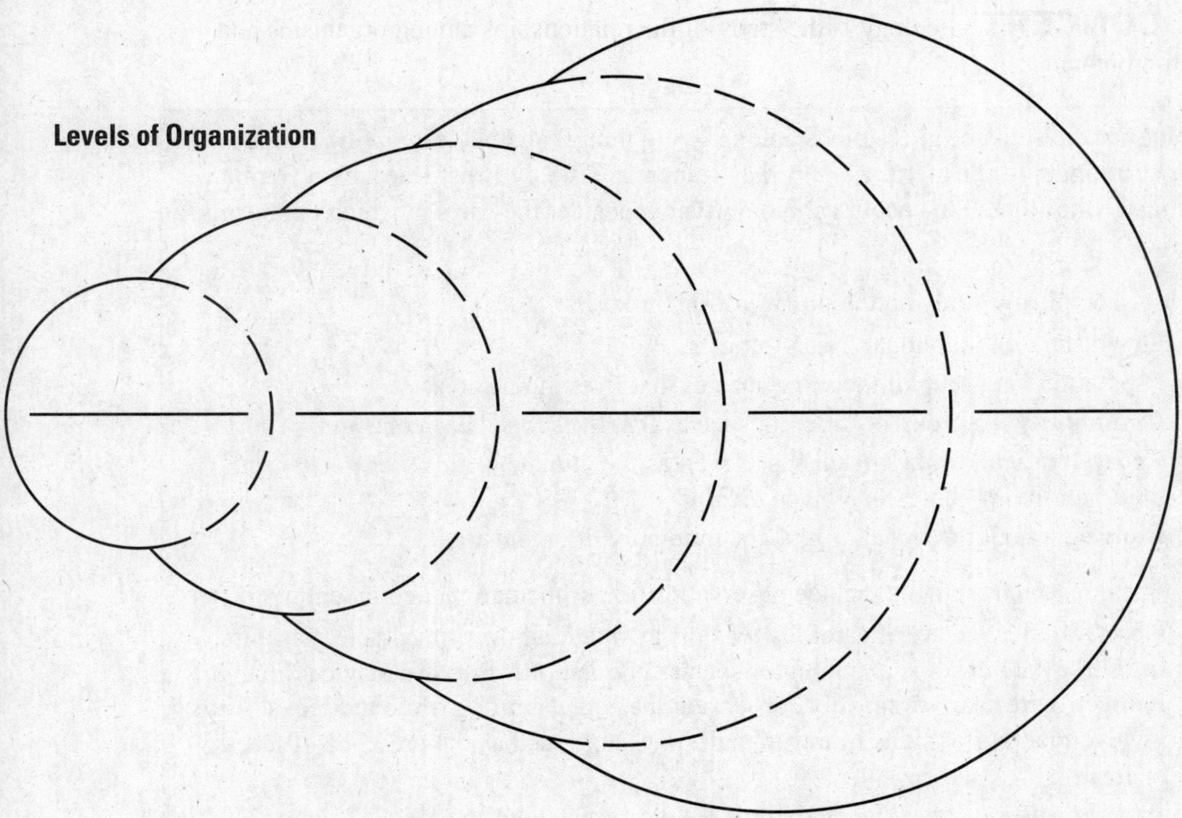

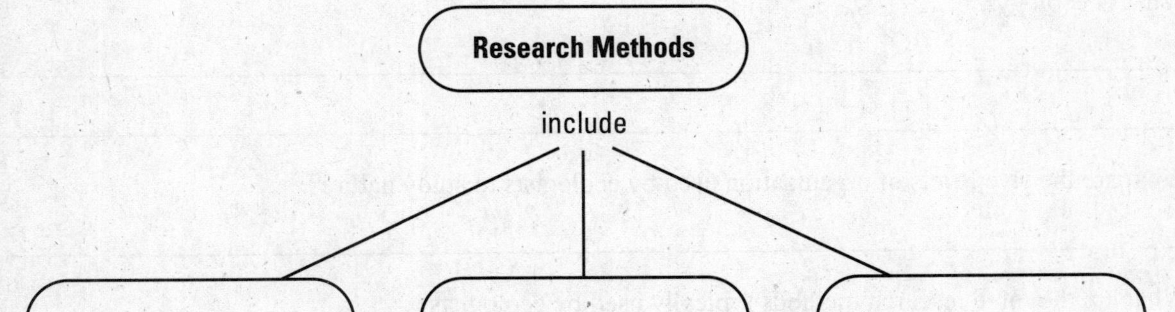

SECTION **13.1** | ECOLOGISTS STUDY RELATIONSHIPS

Reinforcement

KEY CONCEPT Ecology is the study of the relationships among organisms and their environment.

Ecology is the study of interactions among living things, and between living things and their surroundings. The term *ecology* was coined in 1866 by Ernst Haeckel, a German biologist, who wanted to encourage biologists to consider the ways in which organisms interact.

Ecologists typically study nature on five different levels:

- Organism – an individual living thing
- Population – a group of the same species that lives in one area
- **Community** – a group of different species that lives together in one area
- **Ecosystem** – all of the organisms as well as the climate, soil, water, rocks, and other non-living things in a given area
- **Biome** – a major regional or global community of organisms

Ecological research methods include observation, experimentation, and modeling.

- Observation is the act of carefully watching something over time. Scientists often use surveys to observe and monitor species populations. Surveys may be direct or indirect. Direct surveys involve observing the actual animal, while indirect surveys involve looking for signs of the animal's presence, such as the presence of feces or fresh kills.
- Experiments may be conducted in the field or in the lab.
- Models are used when the questions scientists wish to answer cannot be easily answered by observation or experimentation. Models are often used to project what might happen in the future.

1. What is ecology?

____

____

2. What are the five levels of organization used by ecologists to study nature?

____

3. What are the three research methods typically used by ecologists?

____

4. When might a scientist choose to create a model to answer a research question?

____

____

Name Period Date

SECTION
13.2 | BIOTIC AND ABIOTIC FACTORS
Study Guide

KEY CONCEPT
Every ecosystem includes both living and nonliving factors.

VOCABULARY	
biotic	biodiversity
abiotic	keystone species

MAIN IDEA: An ecosystem includes both biotic and abiotic factors.
Use a word from the box below to complete the following sentences.

abiotic	animals	biotic
living	moisture	nonliving
plants	temperature	wind

1. All ecosystems are made up of __________ and __________ components.

2. __________ factors are living things, such as __________ or __________.

3. __________ factors are nonliving things, such as __________, __________, or __________.

MAIN IDEA: Changing one factor in an ecosystem can affect many other factors.

4. Describe what biodiversity means in your own words.

____

5. What is the term for an organism that has an unusually large effect on its ecosystem?

____

6. List a few reasons why a beaver is an example of a keystone species.

____

____

____

Vocabulary Check

7. What is the difference between a biotic and an abiotic factor?

____

____

8. Take another look at the Visual Vocab on page 403. In architecture, a keystone is the stone at the center of an arch that holds the arch together. How does this definition relate to a keystone species?

____

____

Be Creative

In the box below, sketch a simple ecosystem and label the abiotic and biotic factors.

Name Period Date

SECTION **13.2** BIOTIC AND ABIOTIC FACTORS

Power Notes

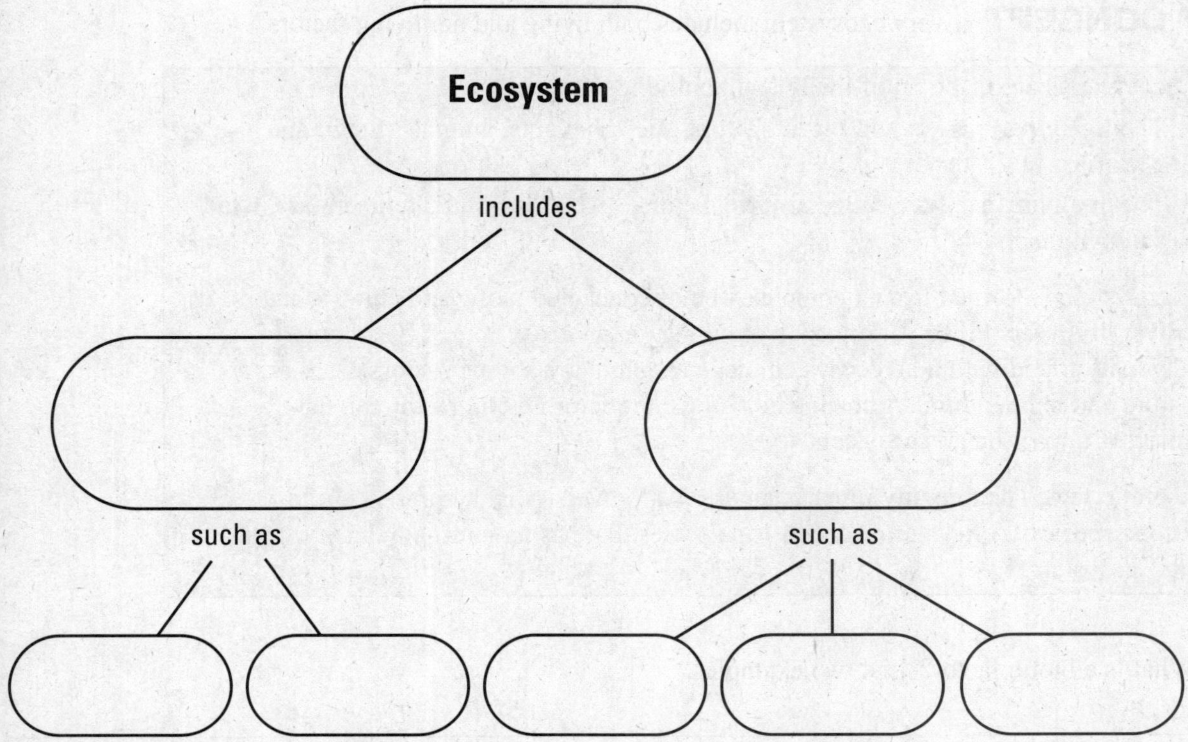

Complex Relationships Within an Ecosystem

Biodiversity is...

A keystone species is...

SECTION 13.2 | BIOTIC AND ABIOTIC FACTORS

Reinforcement

KEY CONCEPT Every ecosystem includes both living and nonliving factors.

All ecosystems are made up of living and nonliving parts.

- The living parts are called **biotic** factors, such as plants, animals, fungi, and bacteria.
- The nonliving parts are called **abiotic** factors, such as moisture, temperature, wind, sunlight, and soil.

An ecosystem is formed from a complex web of connected biotic and abiotic factors. **Biodiversity** refers to the variety of living things in an ecosystem. The amount of biodiversity found within an ecosystem depends on many abiotic factors, such as moisture and temperature. A change in a single biotic or abiotic factor can have a significant impact on an ecosystem.

One biotic factor that greatly impacts an area's biodiversity is the presence of a keystone species. A **keystone species** is a species that has an unusually large effect on its ecosystem.

1. What is a biotic factor? List two examples.

____

____

2. What is an abiotic factor? List two examples.

____

____

3. What is biodiversity?

____

____

4. Why might it be important to preserve areas with high biodiversity?

____

____

____

Name Period Date

SECTION
13.3 | ENERGY IN ECOSYSTEMS
Study Guide

KEY CONCEPT
Life in an ecosystem requires a source of energy.

VOCABULARY	
producer	heterotroph
autotroph	chemosynthesis
consumer	

MAIN IDEA: Producers provide energy for other organisms in an ecosystem.
Complete the following sentences with the correct term.

autotrophs	eating	nonliving
consumers	heterotrophs	producers

1. __________ are organisms that get their energy from __________ resources, meaning they make their own food. These organisms are also called __________.
2. __________ are organisms that get their energy by __________ other organisms. These organisms are also called __________.
3. Why are producers so important to an ecosystem?

____

____

____

4. Why is the Sun important to both producers and consumers?

____

____

____

MAIN IDEA: Almost all producers obtain energy from sunlight.

5. Complete the following Y-diagram to outline the similarities and differences between photosynthesis and chemosynthesis.

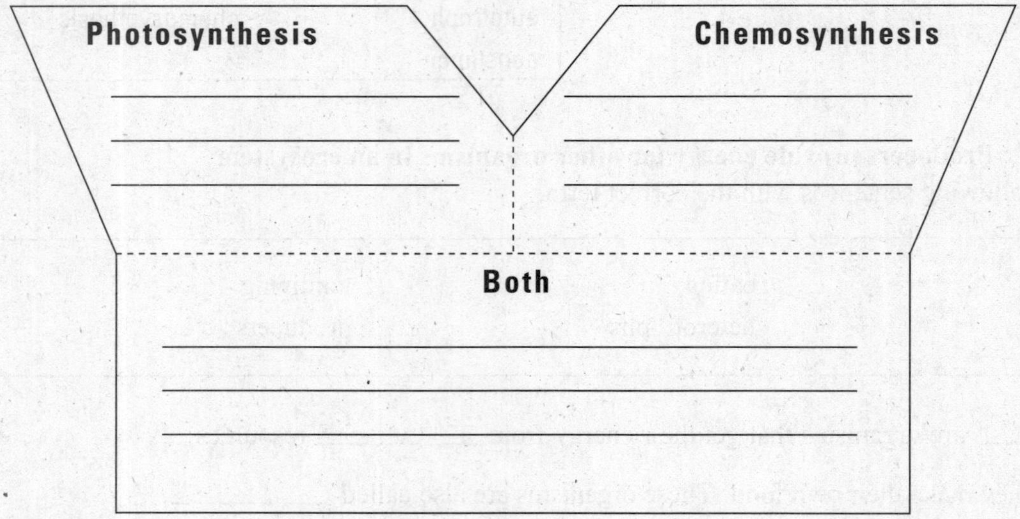

Vocabulary Check

6.

Word Part	Meaning
auto-	self
hetero-	other
-troph	nourishment

Use the above word origins to explain the difference between an autotroph and a heterotroph.

7. The prefix *photo-* means "light" while the prefix *chemo-* means "chemical." How do these word origins relate to the difference between photosynthesis and chemosynthesis?

8. What is the difference between a consumer and a producer?

Name Period Date

SECTION **13.3** | ENERGY IN ECOSYSTEMS

Power Notes

Producers and Consumers

Type	Also Called	Description

Processes by Which Producers Obtain Energy

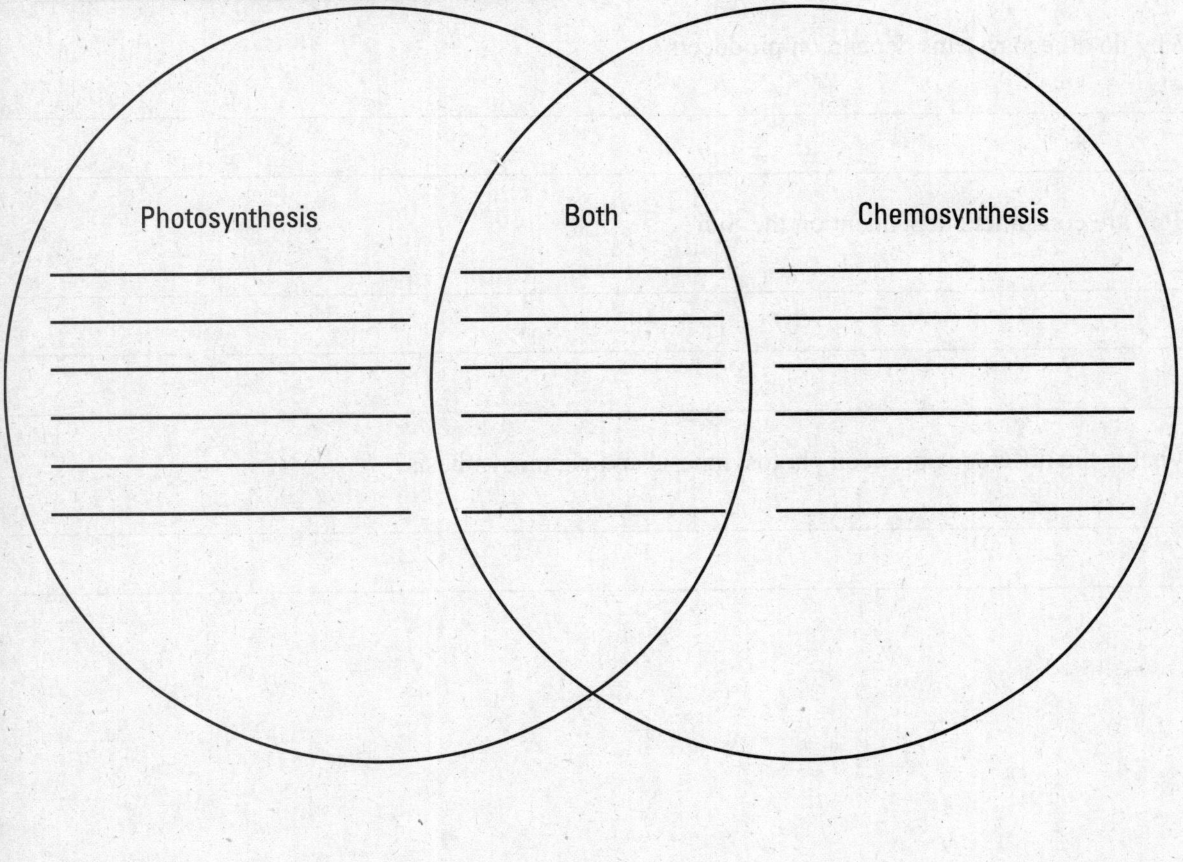

Name Period Date

SECTION **13.3** | ENERGY IN ECOSYSTEMS

Reinforcement

KEY CONCEPT Life in an ecosystem requires a source of energy.

All organisms must have a source of energy in order to survive.

- **Producers** get their energy from nonliving resources, meaning they make their own food. Producers are also called **autotrophs**.
- **Consumers** get their energy by eating other living or once-living resources. Consumers are also called **heterotrophs**.

Photosynthesis is the two-stage process that green plants, cyanobacteria, and some protists use to produce energy in the form of carbohydrates. These chemical reactions form carbohydrates from carbon dioxide and water.

Other producers use chemicals to form carbohydrates in a process called **chemosynthesis.** Chemosynthetic producers are found in deep-sea vent communities as well as in sulfur-rich salt marsh flats and hydrothermal pools.

1. What is the difference between a producer and a consumer?

____

____

____

2. Why do all ecosystems depend on producers?

____

____

3. How are consumers dependent on the Sun?

____

____

____

4. What is the difference between photosynthesis and chemosynthesis?

____

____

Name Period Date

SECTION **13.4** | FOOD CHAINS AND FOOD WEBS

Study Guide

KEY CONCEPT

Food chains and food webs model the flow of energy in an ecosystem.

VOCABULARY	
food chain	decomposer
herbivore	specialist
carnivore	generalist
omnivore	trophic level
detritivore	food web

MAIN IDEA: A food chain is a model that shows a sequence of feeding relationships.

Complete the following sentence with the correct terms.

1. A food chain follows the connection between one ____________ and a single chain of ____________ within an ____________.

Choose the correct term from the box below to fit each description.

carnivore	herbivore	secondary consumer
decomposer	omnivore	tertiary consumer
detritivore	primary consumer	trophic levels

2. I eat only plants. I am a(n) ____________.

3. I eat only other animals. I am a(n) ____________.

4. I eat both plants and animals. I am a(n) ____________.

5. I eat dead organic matter. I am a(n) ____________.

6. I break down organic matter into simpler compounds. I am a(n) ____________.

7. I am the first consumer above the producer level. I am a(n) ____________.

8. I am a carnivore that eats herbivores. I am a(n) ____________.

9. I am a carnivore that eats other carnivores. I am a(n) ____________.

10. The levels of nourishment in a food chain are called ____________.

MAIN IDEA: A food web shows a complex network of feeding relationships.

11. How is a food web different from a food chain?

12. What happens to energy at each link in a food web?

13. What type of organism provides the base of a food web?

Vocabulary Check

14. Use your knowledge of the words *special* and *general* to explain the diets of a specialist and a generalist.

15.

Word Part	Meaning
herba	vegetation
carnus	flesh
omnis	all

Use the word origins to explain the diets of each of the following consumers: herbivores, carnivores, and omnivores.

SECTION **13.4** | FOOD CHAINS AND FOOD WEBS

Power Notes

Types of consumers:

1.

2.

3.

4.

A food web shows:

Add arrows:

Trophic Level
Producer __________ __________
Primary consumers __________ __________ __________ __________
Secondary consumers __________ __________ __________
Tertiary Consumer __________

Name Period Date

SECTION **13.4** | FOOD CHAINS AND FOOD WEBS

Reinforcement

KEY CONCEPT Food chains and food webs model the flow of energy in an ecosystem.

A **food chain** is a model that shows a sequence of feeding relationships between a producer and consumers. There are several types of consumers.

- **Herbivores** eat only plants.
- **Carnivores** eat only animals.
- **Omnivores** eat both plants and animals.
- **Detritivores** eat detritus, or dead organic matter.
- **Decomposers** are detritivores that break down organic matter into simpler compounds.

Food chains are very helpful to explain the feeding relationships of very selective eaters. **Specialists** are consumers that mainly eat only one specific organism or a very small number of organisms. In contrast, **generalists** are consumers that have a varying diet. **Trophic levels** are the levels of nourishment in a food chain. Energy flows up the food chain from the lowest trophic level to the highest.

- Primary consumers (herbivores) are the first consumer above the producer trophic level.
- Secondary consumers (carnivores) eat primary consumers.
- Tertiary consumers (carnivores) eat secondary consumers.

A food web is a model that shows the complex network of feeding relationships and the flow of energy within and sometimes beyond an ecosystem. At each link in a food web, some energy is stored within an organism, and some energy is dissipated into the environment.

1. What are the four main types of consumers?

____

2. What is the difference between a specialist and a generalist?

____

3. What are the trophic levels in a food chain?

____

4. What is the difference between a food chain and a food web?

____

____

5. What happens to energy at each link in a food web?

____

Name Period Date

SECTION **13.5** | CYCLING OF MATTER

Study Guide

KEY CONCEPT

Matter cycles in and out of an ecosystem.

VOCABULARY
hydrologic cycle
biogeochemical cycle
nitrogen fixation

MAIN IDEA: Water cycles through the environment.

Fill in the chart with a description of each process that describes how water moves through an ecosystem in the hydrologic cycle.

Process	Description
1. precipitation	
2. evaporation	
3. transpiration	
4. condensation	

MAIN IDEA: Elements essential for life also cycle through ecosystems.

Complete the following sentences with the proper terms.

5. Plants, animals, and most other organisms need ______________________ for cellular ______________________.

6. Oxygen is released as a waste product by plants during the process of __________________. Animals takes in this oxygen and release it as __________________ during the process of __________________.

7. In the carbon cycle, plants use energy from the Sun to convert __________________ from the air into organic material that becomes a part of the plant's structure.

8. Carbon is released to the atmosphere as carbon dioxide when you breathe during the process of ________________ or through the ________________ of dead organisms.

9. ________________, or the burning of fossil fuels, also adds carbon dioxide to the atmosphere.

10. What is nitrogen fixation?

____

____

11. List five steps that occur during the phosphorus cycle.

____

____

____

____

Vocabulary Check

Use the following word origins to answer the questions below.

Word Part	Meaning
bio-	life
chem-	chemical
geo-	earth
hydro-	water

12. What is a biogeochemical cycle?

____

____

13. What is the hydrologic cycle?

____

____

Name Period Date

SECTION **13.5** | CYCLING OF MATTER

Power Notes

Oxygen cycle:

Carbon cycle:

Hydrologic cycle:

Nitrogen cycle:

Phosphorus cycle:

Name Period Date

SECTION **13.5** | CYCLING OF MATTER

Reinforcement

KEY CONCEPT Matter cycles in and out of an ecosystem.

The **hydrologic cycle** is the circular pathway of water on Earth from the atmosphere, to the surface, below ground, and back. Water falls to Earth as precipitation such as rain or snow. Some droplets of water reenter the atmosphere through evaporation, or from transpiration, which is evaporation that occurs between plant leaves and the atmosphere. Water vapor in the atmosphere condenses and forms clouds, from which precipitation falls.

A **biogeochemical cycle** is the movement of a particular chemical, such as oxygen, carbon, nitrogen, or phosphorus, through the living and nonliving parts of an ecosystem. In the oxygen cycle, oxygen flows into the atmosphere as a byproduct of photosynthesis. Organisms take in this oxygen and release it as carbon dioxide through respiration. Photosynthesis and respiration also figure heavily in the carbon cycle. The cycling of carbon is important because it is the building block of life.

During the nitrogen cycle, bacteria convert gaseous nitrogen into ammonia in a process called **nitrogen fixation.** Nitrifying bacteria change ammonium into nitrate through the process of nitrification. These nitrates are used by plants to make amino acids.

The phosphorus cycle begins when phosphate is released by the erosion of rocks. Plants and fungi can take up the phosphate with their roots. Phosphorus moves from the producers to consumers via the food chain. Phosphorus is returned to the soil through the decomposition of plants and animals. Phosphorus may leach into groundwater from the soil, settling in sediment. Over time this sediment forms into rocks. When these rocks erode, the cycle begins again.

1. What is the hydrologic cycle?

____

____

2. Why is the cycling of elements and nutrients important?

____

____

3. What two major processes are involved in the oxygen cycle?

____

4. What is nitrogen fixation?

____

____

SECTION **13.6** | PYRAMID MODELS
Study Guide

KEY CONCEPT
Pyramids model the distribution of energy and matter in an ecosystem.

VOCABULARY
biomass
energy pyramid

MAIN IDEA: An energy pyramid shows the distribution of energy among trophic levels.
Complete the following sentences with the correct terms.

biomass	heat	waste

1. The measure of the total dry mass of organisms in a given area is called ____________________.

2. When a consumer incorporates the biomass of a producer into its own biomass, a large amount of energy is lost as ____________________ and ____________________.

3. Label the four tiers of the energy pyramid with the correct trophic level (producers, primary consumers, secondary consumers, tertiary consumers).

MAIN IDEA: Other pyramid models illustrate an ecosystem's biomass and distribution of organisms.

Write a description of each pyramid model.

Model	Description
4. energy pyramid	
5. biomass pyramid	
6. pyramid of numbers	

Vocabulary Check

7. What is biomass?

Make an Energy Pyramid

8. Choose an ecosystem. Research what types of plants and animals live in your chosen ecosystem. Draw an energy pyramid that might exist within that ecosystem.

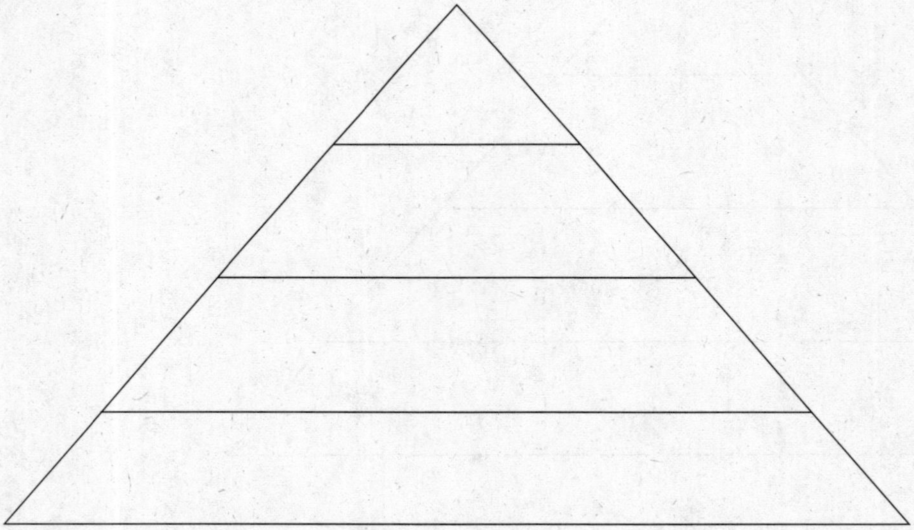

Name Period Date

SECTION
13.6

PYRAMID MODELS

Power Notes

Trophic Levels

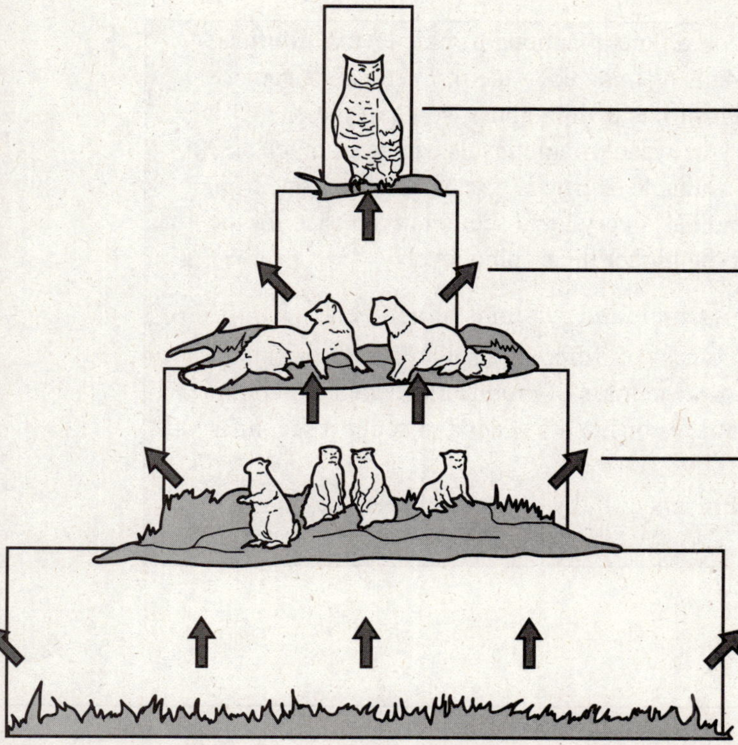

Energy pyramid:

Two other pyramid models:

1. ______________________

Measures:

2. ______________________

Measures:

Name Period Date

SECTION **13.6** | PYRAMID MODELS

Reinforcement

KEY CONCEPT Pyramids model the distribution of energy and matter in an ecosystem.

An **energy pyramid** shows the distribution of energy among trophic levels. **Biomass** is a measure of the total dry mass of organisms in an ecosystem. When a consumer incorporates the biomass from a producer into its own biomass, some of the energy is lost as heat and waste. The loss of energy between trophic levels can be as much as 90 percent, meaning only 10 percent of the available energy is transferred from one trophic level to another. A typical energy pyramid has a very large section at the base for the producers, and tiers that become smaller the higher the trophic level.

Two other pyramid models are biomass pyramids and pyramids of numbers.

- A biomass pyramid compares the biomass of different trophic levels within an ecosystem. This pyramid model shows the mass of producers needed to support primary consumers, the mass of primary consumers needed to support secondary consumers, and so on.
- A pyramid of numbers shows the numbers of individual organisms at each trophic level in an ecosystem.

1. What is an energy pyramid?

____

____

2. What is biomass?

____

____

3. Describe the flow of energy from one trophic level to another.

____

____

4. What is the difference between a biomass pyramid and a pyramid of numbers?

____

____

____

____

Name Period Date

CHAPTER 13

POPULATIONS AND SAMPLES

Data Analysis Practice

Scientists often estimate the size of a large population by sampling the number of individuals in a smaller area and using a formula to calculate the total population number. Once data are collected, a simple equation can be used to find the population estimate.

$T = NA$

T = **Total Population**

N = **Total number of individuals counted / Number of quadrats**

A = **Total area / Area of quadrat**

A scientist wants to estimate the population of sponges on a coral reef using quadrats. Each quadrat is 1 m^2. She counts 450 individuals in 22 quadrats. The total area of the reef is 960 m^2.

1. **Calculate** Use the formula above to calculate the estimated population of sponges.

2. **Infer** After the sampling is completed, an entire section of the coral reef was destroyed by waves caused by a hurricane. How might this affect the validity of the scientist's population estimate?

CHAPTER 13 | DESIGN A MARK-RECAPTURE STUDY

Pre-AP* Activity

**Pre-AP is a registered trademark of the College Board, which was not involved in the production of and does not endorse this product.*

You have learned in Chapter 13 how scientists can use the mark-recapture technique to generate estimates of population size. Whether or not this technique is appropriate or feasible depends upon whether the population is "open" or "closed." The size of an open population—one whose abundance is variable due to birth, death, emigration, or immigration—is difficult to estimate. Closed populations are, by comparison, much easier to assess.

In a closed population, the abundance of organisms is constant for the full duration of the experiment or study. An example of this might be a species of fish in a small pond. Scientists know that the fish aren't going to migrate elsewhere, so they need not worry about the abundance dropping due to emigration. They also know that, barring some unnatural event, fish that are not in the pond at the start of the experiment are not going to spontaneously appear prior to the recapture phase (immigration). Provided that the work is done relatively quickly, the scientists do not have to worry about the effect of predation or other causes of death, either. (If they caught and marked 100 fish in May, but waited until August to recapture them, their estimate of population could be way off due to the loss of many marked fish to predators in the intervening months.) Finally, if the scientists also know that the fish is not going to reproduce in the time between "mark" and "recapture"—or they at least know that any newborn fish will be easily distinguishable from the others—then birth will not be a factor. Thus, the population is closed.

While closed populations do not pose nearly as many challenges as open populations, there are a number of ways that even a simple mark-recapture study can be undermined and the data skewed by poor experimental design.

MARKS

One problem with some mark-recapture studies is the actual method of marking the animals. Some tags or bands used to mark an individual can come off due to a variety of circumstances, from moisture to temperature to the animal's own behavior. It often takes years of trial and error before scientists develop a tag that will actually stay on an organism until someone takes it off. Scientists must also be sure that any mark that is put on an organism does not make that animal more or less likely to be recaptured than an unmarked organism. Moreover, a tag mustn't affect the animal's chance of survival or natural behavior. If, for example, the trauma of capturing and marking an animal results in it spending the duration of the study period in an underground lair, so that the chance of recapture is zero, then the study needs to be redesigned.

Scientists continue to look for less invasive ways to mark or tag organisms. Ideally, an animal may already have a marking that distinguishes it from others. Individual humpback whales, for example, are readily identified by the dark blotches on the white undersides of their tails. These blotches are like huge fingerprints that can be spotted and photographed from many meters away when a whale's tail is held aloft above the sea surface. Scientists can assemble a catalog of photographs of these markings and essentially use them to aid with population estimates and other studies of humpback whale behavior and ecology. Other natural markings on other species can be used the same way.

DESIGN YOUR OWN MARK-RECAPTURE STUDY

In Tanzania, there is a huge crater called Ngorongoro. The floor of the crater is 260 km^2 and the inner walls are 610 m high. The steep angle of these walls makes Ngorongoro a natural enclosure for many of the 25,000 animals living inside, including lions, leopards, elephants, wildebeests, gazelles, and flamingos. In 1994, it was estimated that there were 7000 wildebeests living in the crater.

Answer the following questions on a separate piece of paper.

1. Design a mark-recapture study to get an estimate of what the wildebeest population is now. Consider the following factors: how to mark or visually identify wildebeests, whether or not predation is an issue, the reproductive cycle of the species, and the feasibility of actually capturing or counting the animals.
2. Let's assume that you and your research team mark 500 wildebeests. In the recapture phase, of the 500 animals tallied by you and your team, 50 are marked. In the time between the start of the marking phase and the end of the recapture phase, 2 marked and 18 unmarked wildebeests are observed being killed by lions. What is the ratio of marked to unmarked wildebeests in the recapture data?
3. Compare this ratio to the ratio of marked:unmarked wildebeests killed by lions. What does this suggest about your marking technique?
4. Disregarding predation by lions, how many wildebeests are in the Ngorongoro Crater?
5. If the crater were opened and wildebeests were able to move in and out, what new factors would have to be taken into account in a study of the Ngorongoro Crater wildebeest? Would this population be considered open or closed?

Name Period Date

CHAPTER 13 | BIOMASS IN CORAL REEF ECOSYSTEMS

Pre-AP Activity

One of the most diverse and productive ecosystems on Earth is the coral reef. Scientists estimate that while coral reefs occupy less than one percent of the ocean floor, they account for ten percent of the fish we consume and are home to more than a third of all marine fish species.

THE CORAL REEF FOOD WEB

As you saw in Figure 13.11 on page 410, even a simplified coral reef food web is complicated. Phytoplankton and zooxanthellae—symbiotic algae that live within coral polyps—form the base of the web, converting sunlight into energy that is then taken up, in part, by primary consumers such as sponges, corals, fish, sea turtles, and zooplankton. These primary (1°) consumers are then consumed by secondary (2°) consumers, which are in turn consumed by tertiary consumers (3°), and so on. There are multiple levels, and the variable diets of many species can place them on more than one level. For example, a tiger shark that feeds on a large grouper might in that instance be a quaternary (4°) consumer if the grouper ate a small fish that ate a shrimp that ate some phytoplankton. A few days later, the tiger shark could eat a sea turtle that is a primary consumer, thereby making itself a secondary consumer.

ECOLOGICAL PYRAMIDS

Because of the shifting nature of this complex food web, some coral reef ecologists focus not on individual species but on families or groups of organisms to determine the trophic structure of a reef. For example, scientists might assign all consumers into three groups: primary, secondary, and tertiary. After extensive field work to gather data, the scientists might construct pyramids of biomass or numbers that make it easier to "see" the ecosystem's trophic structure.

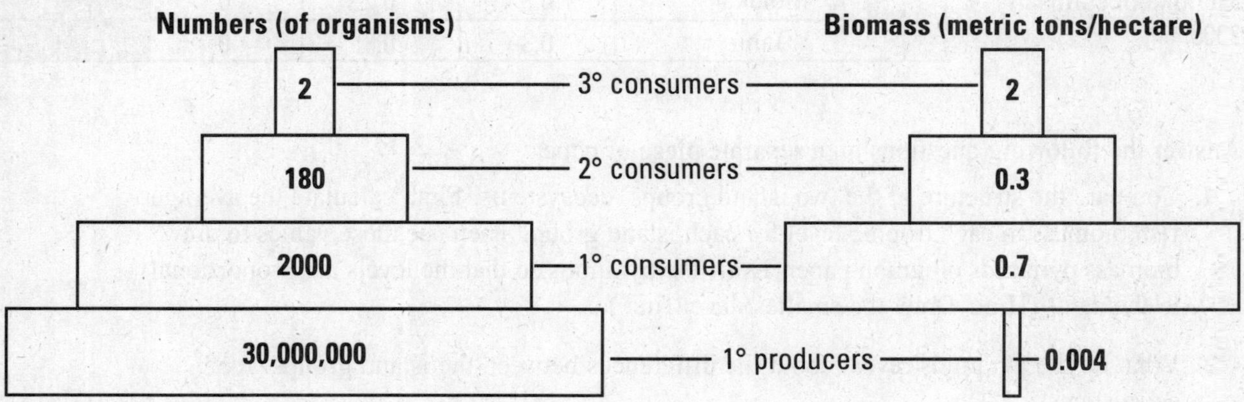

An ecological pyramid represents a hierarchy of feeding relationships in which large numbers or masses of organisms at the base support smaller numbers or masses of organisms above. Pyramids are often drawn in a general, symbolic way. For example, if the pyramid of numbers shown above were drawn proportionally starting from the top level as shown, the drawing of the producer level could end up being larger than your school. And if it were drawn based on the producer level as shown, the top level would be microscopic.

Name Period Date

The shape of a pyramid may also be skewed by the amount of time in which data was collected or the lifespans of the organisms. In the biomass pyramid shown, the producer level is relatively tiny because at any given moment the mass of phytoplankton is small. Because phytoplankton reproduce very quickly, their biomass over time is enough to sustain the consumers. This is analogous to the mass of food in your kitchen. On any given day it is probably less than your family's mass, but over the course of a year—not to mention your lifetime—the mass of food is many times larger than the human biomass in your kitchen "ecosystem."

The table to the right shows data of the average biomass of coral reef fish living around the five Main Hawaiian Islands (MHI) and nine Northwestern Hawaiian Islands (NWHI). The NWHI are home to at least 7000 species, a third of which are endemic, meaning they are found nowhere else on Earth. They were given National Monument status in 2006.

Biomass of Fish in the Two Groups of Hawaiian Islands (metric tons/hectare)

	Island	1° Consumers	2° Consumers	3° Consumers	Total Biomass
NWHI	French Frigate Shoals	0.6	0.3	1.7	2.6
	Gardner	1.6	1	1.3	3.9
	Kure	0.6	0.4	0.3	1.3
	Laysan	0.7	0.2	1.2	2.1
	Lisianski	0.7	0.2	1.8	2.7
	Maro	0.7	0.3	0.7	1.7
	Neckar	0.4	0.3	0.7	1.4
	Nihoa	1.6	0.6	0.6	2.8
	Pearl & Hermes	0.3	0.6	3.8	4.7
MHI	Hawai'i	0.4	0.2	0.1	0.7
	Kauai	0.2	0.2	0	0.4
	Maui	0.5	0.4	0.1	1.0
	Molokai	0.3	0.2	0	0.5
	Oahu	0.3	0.3	0	0.6

Answer the following questions on a separate piece of paper.

1. Compare the structure of the two island groups' ecosystems. First, calculate the average fish biomass in each trophic level for each island group. Then use these values to draw biomass pyramids on graph paper. Draw the pyramids so that the levels are proportional to the data. (Hint: Draw the smallest level first.)

2. What do the pyramids reveal about the differences between the island groups' reef ecosystems?

3. Given that few humans live in the NWHI, what might be responsible for the difference in fish biomass between the two island groups?

4. How can the biomass of 2° consumers support a much larger 3° consumer biomass in the NWHI? Consider the life spans of organisms in these levels, as well as the migratory abilities of top level consumers.

Name Period Date

CHAPTER 13

PRINCIPLES OF ECOLOGY

Vocabulary Practice

ecology	keystone species	herbivore	trophic level
community	producer	carnivore	food web
ecosystem	autotroph	omnivore	hydrologic cycle
biome	consumer	detritivore	biogeochemical cycle
biotic	heterotroph	decomposer	nitrogen fixation
abiotic	chemosynthesis	specialist	biomass
biodiversity	food chain	generalist	energy pyramid

A. Synonyms or Antonyms Identify the words in each pair as synonyms, which are words that mean roughly the same thing, or antonyms, which are words that mean roughly the opposite.

_______________ **1.** producer/autotroph

_______________ **2.** specialist/generalist

_______________ **3.** biotic/abiotic

_______________ **4.** consumer/heterotroph

_______________ **5.** chemosynthesis/photosynthesis

_______________ **6.** herbivore/meat-eater

B. Stepped-Out Vocabulary Define each word. Then write two additional facts that are related to the word.

WORD	DEFINITION	MORE INFORMATION
Example herbivore	*an organism that eats plants*	*primary consumer*
		a cow is an herbivore
1. keystone species		
2. omnivore		
3. decomposer		

VOCABULARY PRACTICE, CONTINUED

C. Word Origins Circle the Greek and Latin word parts in each vocabulary term. Then use the Greek and Latin meanings to construct a very basic definition of the vocabulary word.

bio-	=	life	auto-	=	self	carnus	=	flesh
eco-	=	home	hetero-	=	different	omnis	=	all
syn-	=	together	chemo-	=	chemical	detrere	=	to wear away
-vore	=	eat	photo-	=	light	geo-	=	earth
-troph	=	nourishment	-logy	=	study of			
hydro-	=	water	herba	=	vegetation			

WORD	DEFINITION
1. ecology	
2. photosynthesis	
3. carnivore	
4. herbivore	
5. detritivore	
6. omnivore	
7. chemosynthesis	
8. autotroph	
9. heterotroph	
10. biogeochemical cycle	
11. hydrologic cycle	
12. biomass	

D. Categorize Words Write "A" next to words that can describe abiotic factors. Write "B" next to words that can describe biotic factors.

1. _____ wind _____ sunlight _____ deer

2. _____ soil _____ sunflower _____ water

3. _____ fungus _____ snow _____ eagle

4. _____ temperature _____ prairie dog _____ frog

E. Find the Odd Word Place a check mark next to the word that does not belong and explain why.

1. ____ consumer
____ carnivore
____ plant

Explanation______________________________

2. ____ detritivore
____ producer
____ decomposer

Explanation______________________________

3. ____ omnivore
____ autotroph
____ herbivore

Explanation______________________________

4. ____ trophic level
____ energy pyramid
____ keystone species

Explanation______________________________

F. What's the Difference? For each pair of words below, describe the difference between the two terms.

1. producer/consumer

____

____

2. chemosynthesis/photosynthesis

____

____

3. food chain/food web

____

____

4. community/ecosystem

____

____

Name Period Date

G. Crossword Puzzle Use the clues to solve the puzzle.

Across

1. A diagram that compares energy use among trophic levels

3. Level of nourishment in a food chain

6. Movement of a particular chemical through the living and nonliving parts of an ecosystem

7. All of the organisms as well as the abiotic factors in a given area

8. The variety of living things in an ecosystem

10. A major regional or global community of organisms

Down

2. The process by which gaseous nitrogen is converted into ammonia

4. Detritivore that breaks down organic matter into simpler compounds

5. A consumer that primarily eats one specific organism

6. The measure of the total dry mass of organisms in a given area

7. The study of the interactions among living things, and between living things and their surroundings

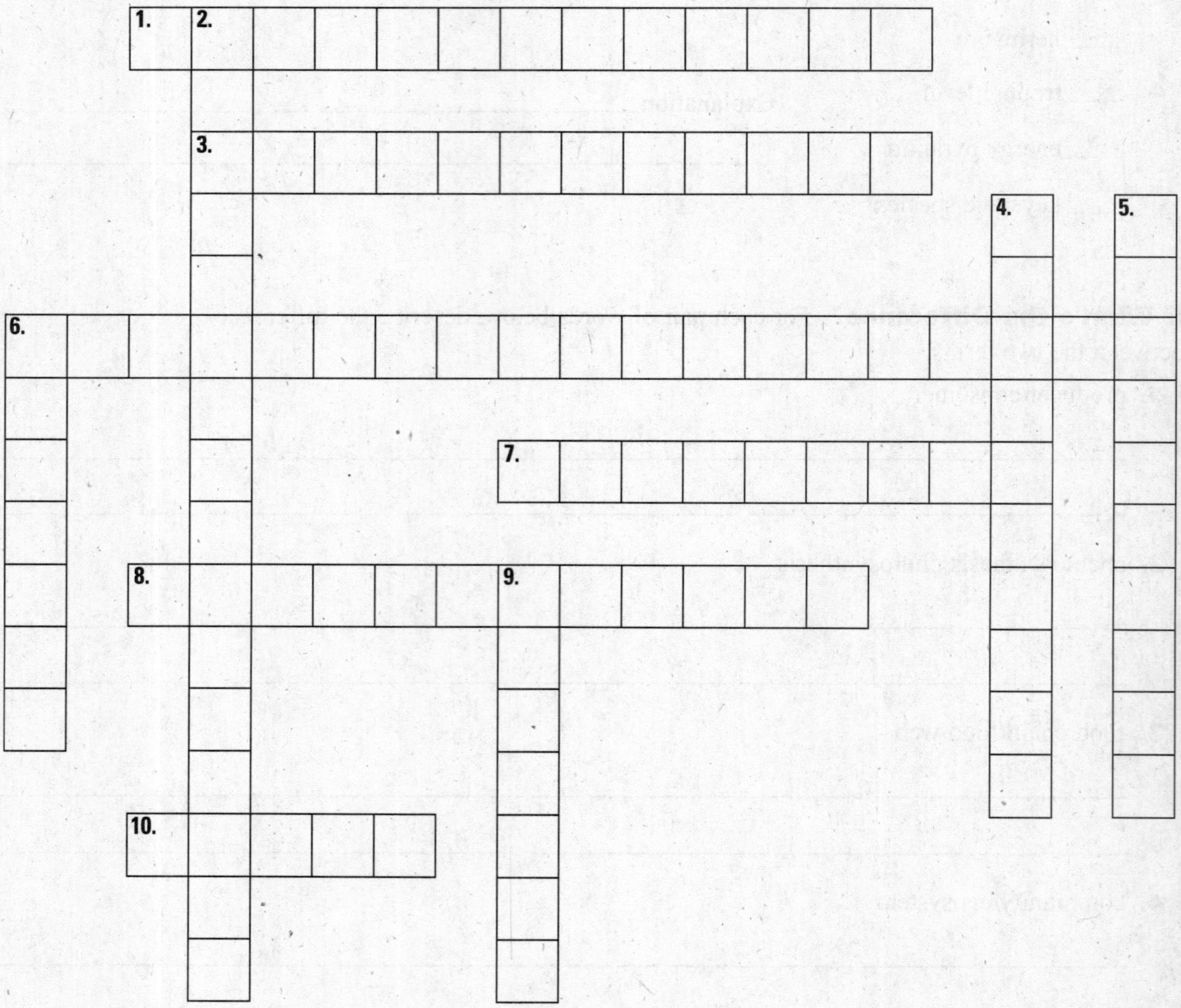

Name Period Date

SECTION **14.1** HABITAT AND NICHE

Study Guide

KEY CONCEPT

Every organism has a habitat and a niche.

VOCABULARY	
habitat	competitive exclusion
ecological niche	ecological equivalent

MAIN IDEA: A habitat differs from a niche.

1. What is the difference between an organism's habitat and its ecological niche?

____

____

2.

food	trees	zebra	grass
hunting behavior	watering hole	sand	savanna
other lions	wildebeest	temperature	

Determine which ecological factors are a part of a lion's niche and which are a part of a lion's habitat by placing the above items in the correct column.

Habitat	Niche

MAIN IDEA: Resource availability gives structure to a community.

3. What is competitive exclusion?

____

____

CHAPTER 14 Interactions in Ecosystems

4. What are the three possible outcomes of competitive exclusion?

5. What are ecological equivalents?

6. Explain why ecological equivalents do not share the same niche.

Vocabulary Check

7. The term *habitat* comes from a Latin word which means "to dwell." Explain how this word origin relates to the definition of a habitat.

8. In competitive exclusion, who is competing and who gets excluded?

9. What does *equivalent* mean in math? How does that meaning relate to ecological equivalents?

Name Period Date

SECTION **14.1** HABITAT AND NICHE

Power Notes

Habitat:

Ecological niche:

Competitive exclusion is a principle that states:

Two other results of competitive exclusion:

-
-

An ecological equivalent is:

Name Period Date

SECTION **14.1** | HABITAT AND NICHE

Reinforcement

KEY CONCEPT Every organism has a habitat and a niche.

A **habitat** is all of the living and nonliving factors in the area where an organism lives. For example, the habitat of a frog includes the water, soil, rocks, sunlight, plants, fish, and other frogs that live in the pond.

A frog also has an ecological niche within its habitat. A frog's **ecological niche** is made up of all the physical, chemical, and biological factors that the frog needs to survive, stay healthy, and reproduce.

- A niche includes factors such as the food the frog eats, the other frogs it competes with for food, and other organisms that may eat the frog.
- Its niche also includes the range of conditions, such as water temperature and oxygen content, that the frog can tolerate.
- A frog's niche includes the way that the frog interacts with other frogs, when it is most active in its habitat, and how it reproduces.

Two different species cannot share the same niche. The principle of **competitive exclusion** states that when two species are competing for the same resources, one species will always be better suited to the niche, and will push out the other species. One of three things will happen:

- One species will go extinct.
- The resources of the niche will be divided and the species will coexist.
- An evolutionary response will result in selection of different traits that are successful in different parts of the niche.

In different communities, ecological equivalents may have very similar niches. **Ecological equivalents** are species that occupy similar niches but live in different geographical regions.

1. How is a habitat different from a niche?

____

____

____

2. What are the possible outcomes of competitive exclusion?

____

____

3. How can ecological equivalents occur?

____

Name Period Date

SECTION **14.2** | COMMUNITY INTERACTIONS

Study Guide

KEY CONCEPT
Organisms interact as individuals and as populations.

VOCABULARY		
competition	symbiosis	commensalism
predation	mutualism	parasitism

MAIN IDEA: Competition and predation are two important ways in which organisms interact.

Next to each situation described below, write whether it is an example of *inter*specific competition or *intra*specific competition.

_______________ **1.** Two squirrels race up a tree to reach a hidden pile of nuts.

_______________ **2.** A hyena chases off a vulture to feast on an antelope carcass.

_______________ **3.** Different species of shrubs and grasses on the forest floor compete for sunlight.

_______________ **4.** Brown bears hunting for fish on a river's edge fight over space.

_______________ **5.** Male big horn sheep butt heads violently in competition for mates.

6. Draw and label a sketch that represents an example of a predator-prey interaction.

CHAPTER 14 Interactions in Ecosystems

MAIN IDEA: Symbiosis is a close relationship between species.

7. For each type of symbiotic relationship, complete the chart with details about how each organism is impacted using the terms "Benefits," "Harmed," or "No impact." For each situation, assume that Organism A initiates the relationship.

Symbiotic Relationship	Organism A	Organism B
mutualism		
commensalism		
parasitism		

8. How is parasitism similar to and different from predation?

9. What is the difference between endoparasites and ectoparasites?

Vocabulary Check

10. The term *symbiosis* comes from a Greek term which means "living together." How does this word origin help to explain the definition of symbiosis?

11. Use your knowledge of the word "mutual" to write a definition for mutualism.

12. The word *commensalism* comes from the Latin *mensa*, meaning "table," and *com-*, meaning "with." If I come to your table to eat your food, I benefit but you don't. Draw a sketch to show this meaning to help you remember it.

Name Period Date

SECTION
14.2 COMMUNITY INTERACTIONS
Power Notes

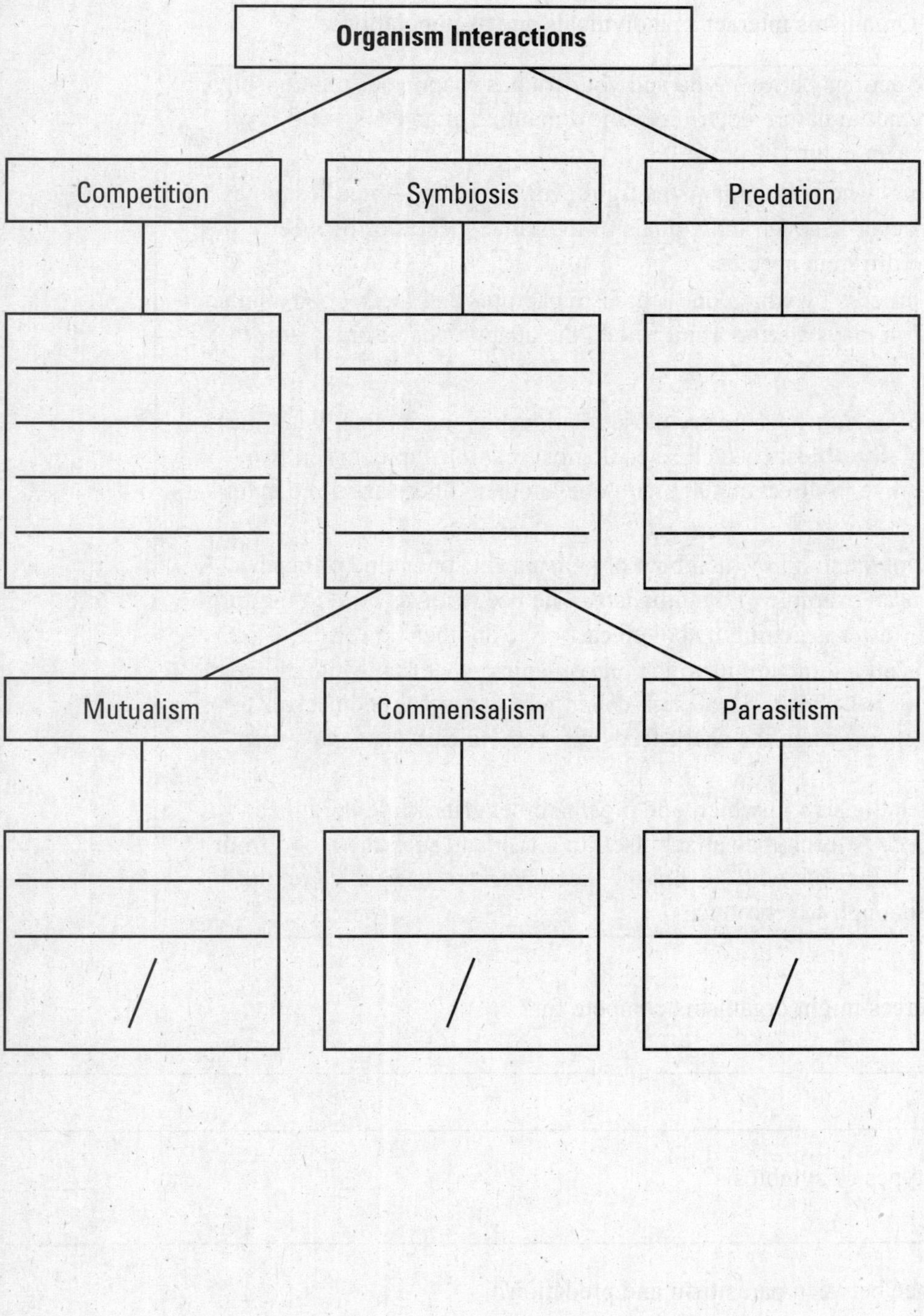

Name Period Date

SECTION 14.2 | COMMUNITY INTERACTIONS

Reinforcement

KEY CONCEPT Organisms interact as individuals and as populations.

Similar to how the interactions between you and your friends shape your relationships, the way organisms interact in nature determines the dynamics of an ecosystem. Two major interactions occur in nature:

- **Competition** occurs when two organisms fight over the same limited resources. Competition can occur between individuals of the same species or between individuals of two different species.
- **Predation** is the process by which one organism captures and feeds upon another organism. Predation plays an important role in the adaptations of organisms to their habitat.

In some cases, two species may have a very close relationship and interact with one another very frequently. **Symbiosis** is a close ecological relationship between two or more organisms that live in direct contact with one another. There are three main types of symbiosis:

- **Mutualism** is an interaction in which both organisms get some kind of benefit. A bee and a flower is an example of a mutualism. The bee receives food in the form of nectar, and the flower is getting its pollen carried to another flower.
- **Commensalism** is an interaction in which one organism benefits, while the other neither benefits nor is harmed. Small fish called remoras attach themselves to the sides of sharks, and when the shark feeds, the remora eats the scraps the shark cannot eat.
- **Parasitism** is a relationship in which one organism benefits while the other organism is harmed. A leech may attach itself to a fish and suck the blood from the fish. Eventually the fish will die, but the leech has kept itself alive on the fish's blood long enough to reproduce.

1. What types of resources might organisms compete for?

____

____

2. What are the three types of symbiosis?

____

3. What is the difference between parasitism and predation?

____

____

____

Name Period Date

SECTION **14.3** | POPULATION DENSITY AND DISTRIBUTION

Study Guide

KEY CONCEPT

Each population has a density, a dispersion, and a reproductive strategy.

VOCABULARY	
population density	survivorship curve
population dispersion	

MAIN IDEA: Population density is the number of individuals that live in a defined area.

1. What is the formula for calculating population density?

____

2. What might cause the population density of a population of deer to increase?

____

MAIN IDEA: Geographic dispersion of a population shows how individuals in a population are spaced.

3. In the boxes below, draw and label the three types of population dispersion patterns.

_______________	_______________	_______________

4. List two reasons why a population might live in a clumped dispersion and two reasons why a population might live in a uniform dispersion.

____

____

CHAPTER 14 Interactions in Ecosystems

MAIN IDEA: Survivorship curves help to describe the reproductive strategy of a species.

5. What is meant by the term *reproductive strategy*? What accounts for differences in reproductive strategies?

____

____

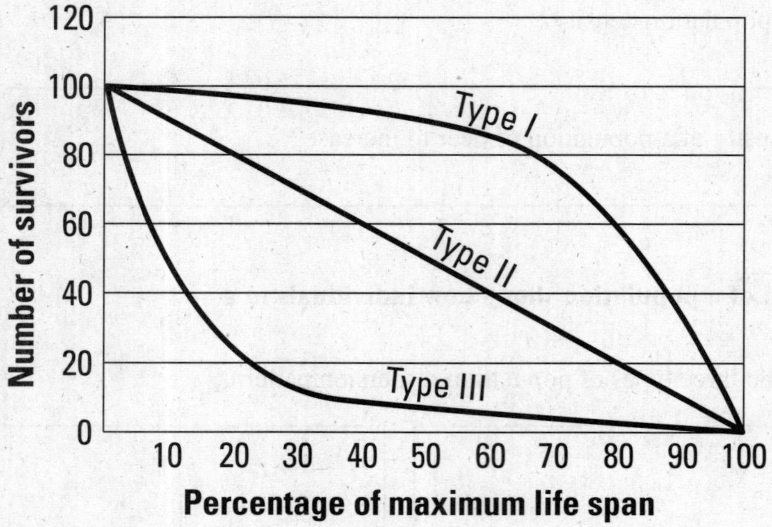

Take a look at each of the survivorship curves shown above. Next to each type of organism listed below, write in the space provided whether it is an example of Type I, Type II, or Type III survivorship.

_______________ **6.** lion

_______________ **7.** bird

_______________ **8.** reptile

_______________ **9.** small mammal

_______________ **10.** invertebrate

_______________ **11.** fish

_______________ **12.** giraffe

_______________ **13.** human

Vocabulary Check

14. What is the difference between population density and population dispersion?

____

____

Name Period Date

SECTION
14.3
POPULATION DENSITY AND DISTRIBUTION
Power Notes

Population density is:

Calculated Using the Formula:

________________ =

Population dispersion is:

Three dispersion types:

__________ __________ __________

A survivorship curve is:

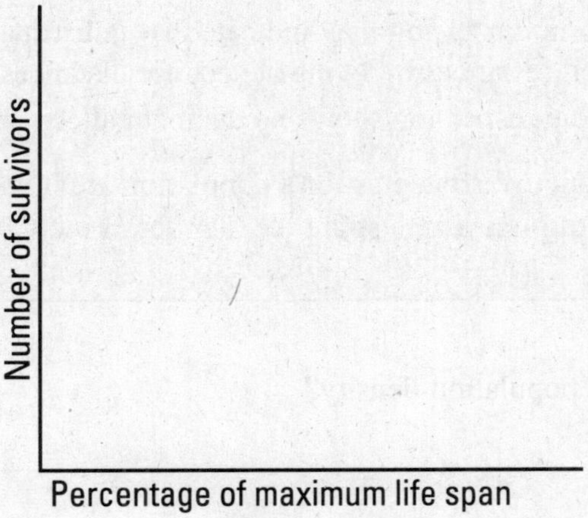

Three Types of Survivorship Curves

Type	Description
Type I	• •
Type II	• •
Type III	• •

Name Period Date

SECTION
14.3 POPULATION DENSITY AND DISTRIBUTION

Reinforcement

KEY CONCEPT Each population has a density, a dispersion, and a reproductive strategy.

Recall that a population is a group of the same species living in the same area. A population can be measured in many ways. One way is by its density. **Population density** is a measure of the number of individuals living in a defined area. Population density is measured by creating a ratio of individuals that live in a particular area to the size of that particular area. The formula for population density is

# of individuals / area (units²) = population density

For example, if there are 50 deer living in an area of 10 km², the population density would be 5 deer per km².

A population can also have a dispersion pattern. **Population dispersion** is how the individuals of a population are spread out in a specific area. There are three types of population dispersion patterns:

- Clumped dispersion shows that individuals live close together in groups or packs. This type of dispersion may help with hunting and feeding, as well as protection from predators.
- Uniform dispersion may indicate that individuals are territorial and compete for limited resources by living at specific distances from one another.
- Random dispersion shows no distinct pattern within a specific area.

The reproductive strategies for a population are illustrated through survivorship curves. **Survivorship curves** illustrate the number of individuals in a population surviving over time.

1. What is population density?

____

____

2. Calculate the population density for a group of 30 birds that live in an area of 3 km².

____

3. What are the three types of population dispersion patterns and what are the characteristics of each population?

____

____

____

____

Name Period Date

SECTION **14.4** | POPULATION GROWTH PATTERNS

Study Guide

KEY CONCEPT
Populations grow in predictable patterns.

VOCABULARY		
immigration	logistic growth	limiting factor
emigration	carrying capacity	density-dependent limiting factor
exponential growth	population crash	density-independent limiting factor

MAIN IDEA: Changes in a population's size are determined by immigration, births, emigration, and deaths.

Choose a word from the box below that best completes each sentence.

births	emigration	deaths	immigration

1. When resources are abundant in a particular area, individuals may move into the population of this area. This movement of individuals into a population from a different population is called ________________.

2. A very cold winter has left many deer in a population hungry and sick. By the end of the winter, this population will likely decrease because of ________________.

3. A deer population experiences growth when the rate of reproduction increases. This change in population size is due to ________________.

4. As humans move into their territory, many members of a deer population move away and join other herds. This movement of individuals out of a population into a new population is called ________________.

5. How does the availability of resources affect population growth?

____

____

CHAPTER 14 Interactions in Ecosystems

MAIN IDEA: Population growth is based on available resources.

6. In the space below, draw and label the two different types of population growth curves. Write a brief description next to each graph.

7. What type of population growth curve shows a carrying capacity?

8. What type of population growth is at risk for a population crash? Explain why.

MAIN IDEA: Ecological factors limit population growth.

8. List three examples of density-dependent limiting factors.

9. List three examples of density-independent limiting factors.

Vocabulary Check

Explain why each pair of words below are opposites.

10. emigrate/immigrate

11. density-dependent limiting factor/density-independent limiting factor

12. exponential growth/logistic growth

Name Period Date

SECTION
14.4 POPULATION GROWTH PATTERNS
Power Notes

Four factors that affect the size of a population:

- ______________________
- ______________________
- ______________________
- ______________________

Exponential Growth

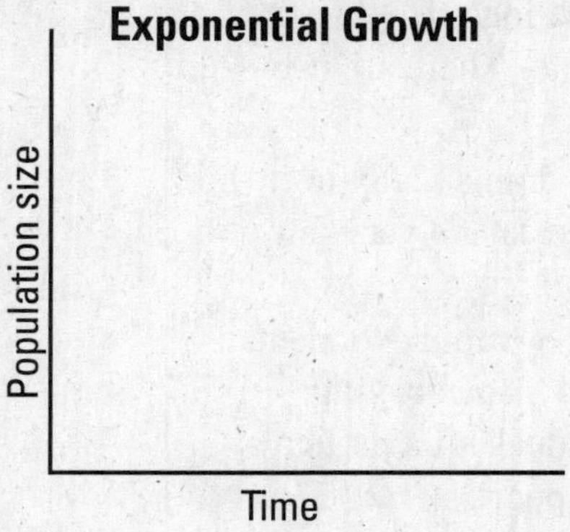

Logistic Growth

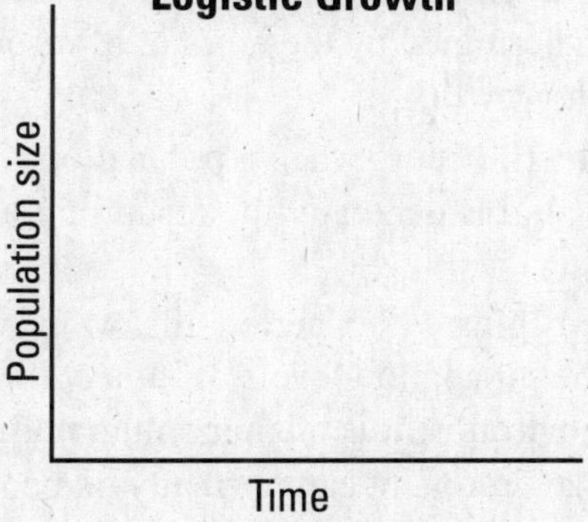

Density-dependent limiting factors:

- ______________________
- ______________________
- ______________________

Density-independent limiting factors:

- ______________________
- ______________________
- ______________________

Name Period Date

SECTION **14.4** | POPULATION GROWTH PATTERNS

Reinforcement

KEY CONCEPT Populations grow in predictable patterns.

Over time, the size of a population increases and decreases. These changes are due to four factors:

- **Immigration** is the movement of individuals into a population from another population and increases the size of a population.
- Births occur when individuals in a population reproduce and result in an increase in population size.
- **Emigration** is the movement of individuals out of a population and into another population and results in a decrease in population size.
- Deaths occur when predation, disease, or old age decrease the size of a population.

The growth of a population is a function of the environmental conditions. How fast a population grows is determined by the amount of resources available. There are two patterns of population growth:

- **Exponential growth** occurs when a population size increases dramatically over a period of time, and is generally the result of abundant resources and very low levels of predation.
- **Logistic growth** begins with a period of slow growth followed by rapid exponential growth before the population levels off at a carrying capacity. The **carrying capacity** of an environment is the maximum number of individuals of a particular species that the environment can normally and consistently support.

Population sizes are kept in check by limiting factors. A **limiting factor** is any environmental influence that directly affects a population size. **Density-dependent limiting factors** are affected by the number of individuals living in a given area. They include competition, predation, and disease. **Density-independent limiting factors** are factors that limit the growth of a population regardless of its density. These factors include unusual weather, natural disasters, and human activities.

1. What four factors influence the size of a population?

____

2. What is carrying capacity? What type of population growth does it affect?

____

____

3. What is the difference between a density-dependent limiting factor and a density-independent limiting factor?

____

____

Name Period Date

SECTION
14.5

ECOLOGICAL SUCCESSION
Study Guide

KEY CONCEPT
Ecological succession is a process of change in the species that make up a community.

VOCABULARY	
succession	pioneer species
primary succession	secondary succession

MAIN IDEA: Succession occurs following a disturbance in an ecosystem.

1. What is ecological succession?

____

____

2. Fill in the chart below with a description and simple sketch of the four main steps of primary succession. Include the amount of time it takes for each stage of this process.

3. Fill in the chart below with a description and simple sketch of the four main steps of secondary succession. Include the amount of time it takes for each stage of this process.

Vocabulary Check

4. What is the difference between primary and secondary succession?

5. Use your knowledge of the word *pioneer* to write a definition for the term *pioneer species*.

Name Period Date

SECTION **14.5** | ECOLOGICAL SUCCESSION

Power Notes

Primary succession is: ____

____

________	________	________	________

Secondary succession is: ____

____

________	________	________	________

CHAPTER 14
Interactions in Ecosystems

Name Period Date

SECTION **14.5** | ECOLOGICAL SUCCESSION
Reinforcement

KEY CONCEPT Ecological succession is a process of change in the species that make up a community.

Each time an ecosystem is damaged, the process of succession re-forms the area. **Succession** is the sequence of biotic changes that regenerate a damaged community or create a community in a previously uninhabited area. Succession is a process with no distinct beginning or end. In a community, succession is always occurring.

After a volcano erupts, the molten lava hardens and leaves behind nothing but solid rock. **Primary succession** is a type of succession that begins with a previously uninhabited, barren landscape. **Pioneer species** are the first organisms that live in this type of habitat. Pioneer species begin the process of breaking down the rock into soil that can hold plants. This process may take hundreds of years, but eventually the soil produced by pioneer species will give rise to entire ecosystems of plants, animals, and other organisms.

More often an environment had many different plants and animals, but a disaster such as a fire or flood may have destroyed much of the habitat. **Secondary succession** is the reestablishment of a damaged ecosystem in an area where the soil was left intact. The dynamic processes of succession are always changing the face of an ecosystem.

1. What is succession?

____

____

2. Why are pioneer species so important for primary succession?

____

____

____

3. Explain why succession is a never-ending process.

____

____

____

Name Period Date

CHAPTER 14 | READING A COMBINATION GRAPH

Data Analysis Practice

Combination graphs show two or more sets of data on the same graph.

Scientists have been tracking the population numbers of snowshoe hares, lynx, and coyotes in northern Canada over many years. In this region, lynx and coyote are the primary predators of the snowshoe hare. The graph below shows the population numbers for all three animals over a ten-year period.

GRAPH 1. LYNX, COYOTE, AND HARE POPULATIONS

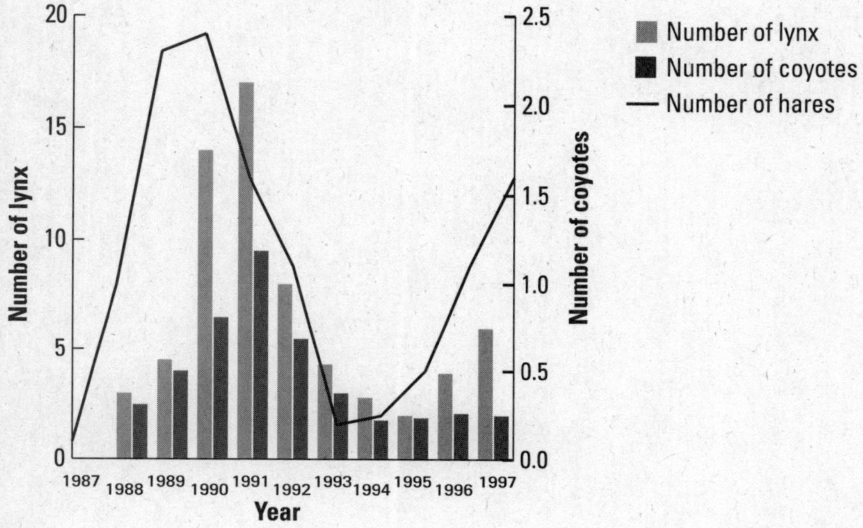

1. **Analyze** As the population of snowshoe hares increases, what happens to the coyote and lynx populations?

____

____

____

2. **Identity** Scientists have observed that the population of hares follows a pattern that occurs in an eight-year cycle. Over this time period, the population peaks and then crashes. Predict how snowshoe hare, lynx, and coyote populations would change by extending the graph for the years 1998–2003.

____

____

____

Name [illegible] Period [illegible] Date [illegible]

SKILL: READING A COMBINED GRAPH

14 Data Analysis Practice

Combination graphs show two or more sets of data on the same graph.

Scientists have been studying the population numbers of [illegible] there [illegible] in this region. [illegible] are the primary predators of [illegible]. The graph below shows the population counts [illegible] over a [illegible]-year period.

GRAPH 1. [illegible] POPULATIONS

1. **Analyze** As the population of [illegible] increases, what happens to the [illegible] and [illegible]?

2. **Hypothesize** Scientists have observed that the population of [illegible] [illegible]. During this time period, the population [illegible] [illegible] and [illegible] populations [illegible] 1988–2003 [illegible]

Name Period Date

CHAPTER 14 | BROOD PARASITISM

Pre-AP Activity

You have learned in Chapter 14 that parasitism is a type of symbiosis in which one species causes harm to another. When we talk about parasites, most of us think of endoparasites—organisms that live and feed inside a host's body—or ectoparasites, which feed on the outside of a host. However, some forms of parasitism are behavioral rather than physiological.

BROOD PARASITISM

Brood parasites are birds that lay their eggs in the nests of other bird species. This frees the parasitic bird from investing energy in building a nest and raising its own young. Instead, the parasite can use that energy to produce more eggs. The effect on the host species is negative. In most cases, the host will incubate the parasite's eggs and rear its nestlings as she would her own. This effectively creates another mouth to feed, which can result in malnutrition and death of her own nestlings.

Brood parasitism has been observed in a range of species, such as the black-headed duck, African honeyguides, and European cuckoos. In North America, the best studied and perhaps most notorious brood parasite is the brown-headed cowbird, *Molothrus ater.* Some brood parasite species are nonobligate parasites: under some circumstances, they will incubate and rear their own young. Brown-headed cowbirds are obligate parasites—they must parasitize nests in order to successfully reproduce. Researchers think that over time the cowbird lost the ability to build nests and incubate young. As a result, it is completely dependent on the host to ensure its own reproductive success.

The brown-headed cowbird is found in most parts of North America below the Arctic Circle. It favors edge habitats—where open areas such as meadows meet woodlands. The cowbird got its name from the relationship it had several hundred years ago. Cowbirds lived in close proximity to the bison of the Great Plains, feeding on insects and grass seeds kicked up by the bison as they trudged along. As the bison population collapsed due to hunting, cowbirds adapted by forming similar associations with cattle and other livestock. Over the last several decades, cowbirds have expanded their range into suburban and city areas. This largely has been due to wide-scale habitat fragmentation: clearing portions of woodlands for development has created an abundance of edge habitats.

HOW BROOD PARASITISM WORKS

Before she begins to lay eggs, a female brown-headed cowbird will monitor potential hosts. Very early in the morning, before dawn, she will watch other birds as they build nests. When the birds fly off to get more nest material, the cowbird will fly in and check the nest. If it is ready, she will deposit one egg and then leave. If the host bird has already laid its own eggs, the cowbird will often remove one of those eggs before flying off. While many brood parasites produce eggs that mimic those of their hosts, the brown-headed cowbird does not; however, when host birds return to their nests, they generally do not notice that there is an additional egg there.

Within two weeks, the cowbird nestling is born, usually before the host's own offspring hatch. The cowbird is then fed and reared by the host bird. Cowbirds tend to outcompete the host's own offspring for food, in part because they hatch earlier. This lowers the reproductive success of the host. Some host species, such as robins, are able to recognize cowbird eggs right away. They can either destroy them in the nest or throw them out. Other species simply abandon a nest if they find that a cowbird has parasitized it.

THE CONSERVATION CONNECTION

Biologists have documented that the brown-headed cowbird parasitizes more than 200 host species, many of which are migratory songbirds. Over the three-month breeding season, a female cowbird may lay as many as 40 eggs. Because she lays only one egg per nest, this means that a single female parasitizes up to forty nests per year, and potentially lowers the reproductive success of forty other female birds.

The great majority of the cowbird's victims are songbirds. In general, songbird populations have decreased in recent years. Habitat loss is a key element of this decline. Most North American songbirds spend the winter in Florida or the Caribbean, and when they arrive home each year for breeding, their former habitat often has been destroyed or altered by suburban developments. The prevalence of cowbird parasitism on songbirds is another factor in the decline of songbirds. Roughly 90 percent of the nests of some songbird species, such as the black-capped vireo, are parasitized each year.

Answer the following questions on a separate piece of paper.

1. Given that the brown-headed cowbird is totally dependent on the nests of other birds, how might the decline of songbird populations affect cowbird populations? And how might this effect on cowbirds end up affecting the songbirds?

2. Imagine that you are an ornithologist specializing in the ecology of brood parasites. You hypothesize that robins are more likely to remove or destroy cowbird eggs laid in their nests if they have not yet laid any of their own eggs. Design an experiment or study to test this hypothesis in a nearby habitat known for its songbird and cowbird populations. What types of data would you collect and compare?

3. Suppose that your study disproves your hypothesis, and robins are actually more likely to destroy cowbird eggs if they are laid after their own eggs have been laid. What might this suggest about robins, and in particular, what might it say about their sensory capabilities?

Name Period Date

CHAPTER 14 | THE EFFECTS OF POPULATION DENSITY AND CLIMATE ON PLAGUE

Pre-AP Activity

You have learned in Chapter 14 that population size can be regulated by density-independent and density-dependent factors. As you will discover, often there are numerous factors affecting a population all at once, or in sequence. A given population might be affected by a density-dependent factor, such as disease and a density-independent factor, such as temperature or rainfall, at the same time. For example, as the number of squirrels in a hypothetical population increases, the population becomes overcrowded, and a pathogen could be transmitted more readily. More squirrels die or become too ill to reproduce, leading to a decrease in population size. In this example, infectious disease is a density-dependent factor for the squirrels. But what else is going on in this example? How did the squirrel population increase in the first place? Did it grow quickly or gradually? Was a density-independent factor, such as climatic conditions, involved?

PLAGUE: AN OVERVIEW

Plague has been responsible for some of the deadliest epidemics in recorded history, including the Black Death, which claimed 24 million lives between 1346 and 1352. It remains an active threat to public health, with outbreaks occurring in several parts of the world almost annually. The disease is caused by infection with the bacterium *Yersinia pestis*. In nature this bacterium cycles between certain wild rodent species and their fleas. This is sometimes called the sylvatic cycle. When human populations settle near rural areas, the bacterium may enter what some scientists call the urban cycle, in which infected fleas infest urban rodents, domestic animals, and humans. It also can be transmitted through direct contact with an infected animal or its tissues. These forms of transmission cause the bubonic form of plague. Symptoms include high fever, chills, fatigue, body aches, and painful swellings of lymph nodes in the groin or the neck. Another form of plague, pneumonic plague, occurs when the bacterium invades the respiratory system. When the host coughs or sneezes, infectious droplets that are expelled into the air can then be inhaled by other hosts. The third form, septicemic, results when the bacterium directly enters the host's circulation, usually through direct contact with infected tissues.

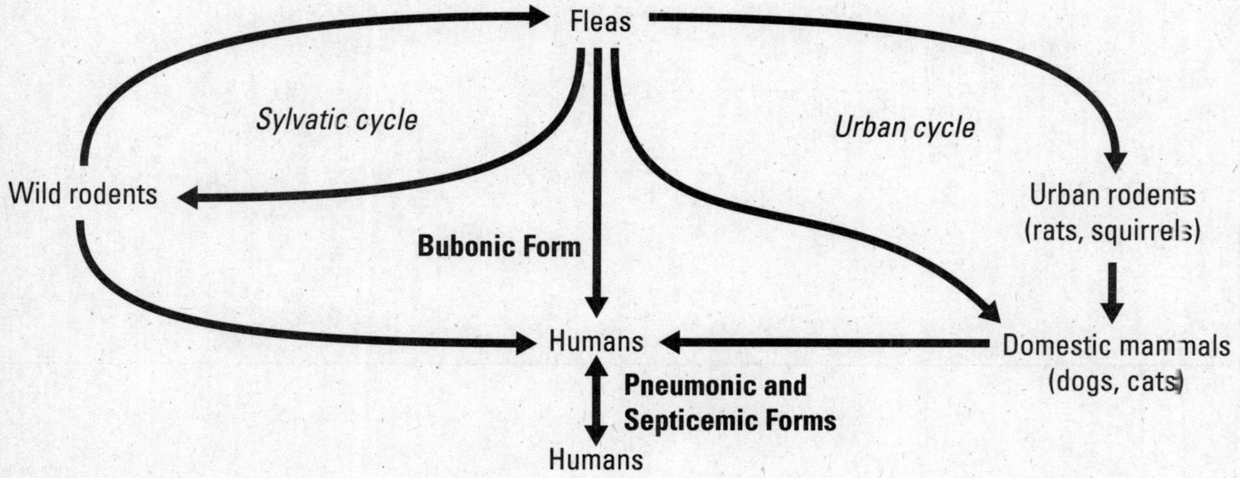

CHAPTER 14 Interactions in Ecosystems

Name Period Date

ENVIRONMENTAL FACTORS DRIVING PLAGUE

Independent studies of plague outbreaks in wild rodents in Central Asia and New Mexico have shown a connection between climate, host population density, and disease outbreaks. In New Mexico, plague is endemic in desert rodents. A group of scientists noticed an increased incidence of plague outbreaks following unusually wet spring seasons. The scientists examined almost 50 years of weather data and plague reports, and found that every outbreak of plague had been preceded by an unusually wet spring. Further examination of the data revealed that rodent populations increased dramatically following high rainfall.

Halfway around the world, a similar set of observations was made by scientists studying plague outbreaks in Kazakhstan, where the major host of the disease is the great gerbil. The researchers noted that plague outbreaks always occurred roughly two years after the great gerbil population reached a critical number. If the population was smaller than the critical number, major outbreaks did not occur. Having noted a connection between climate shifts and plague outbreaks, the scientists next examined archived climate and outbreak data. Just like in New Mexico, the scientists found a distinct relationship between climate and plague. Outbreaks consistently occurred two years after an unusually warm spring or wet summer. Since the earlier study showed that outbreaks only occurred when the population had expanded, the climate data suggested that climate changes led to rapid population growth that allowed the gerbil population to reach a critical number that preceded plague outbreaks.

YOUR TURN

Construct a sequence diagram that shows how increased rainfall and warmer temperatures might increase host population density and ultimately produce a plague outbreak. Think about the issue on a large scale. What effect would warmer temperatures and increased rainfall have on the environmental conditions of a wild rodent population?

Name Period Date

CHAPTER 14 | INTERACTIONS IN ECOSYSTEMS

Vocabulary Practice

habitat	parasitism	population crash
ecological niche	population density	limiting factor
competitive exclusion	population dispersion	density-dependent limiting factor
ecological equivalent	survivorship curve	density-independent limiting factor
competition	immigration	succession
predation	emigration	primary succession
symbiosis	exponential growth	pioneer species
mutualism	logistic growth	secondary succession
commensalism	carrying capacity	

A. What's the Difference? For each pair of words below, describe the difference between the two terms.

1. primary succession/secondary succession

____

____

2. ecological niche/habitat

____

____

3. logistic growth/exponential growth

____

____

4. density-dependent limiting factor/density-independent limiting factor

____

____

5. mutualism/parasitism

____

____

VOCABULARY PRACTICE, CONTINUED

B. Matching Write the vocabulary term next to its definition.

commensalism	mutualism	predation
competition	parasitism	symbiosis

_______________ **1.** A close relationship between two or more individuals of different species that live in close contact with one another

_______________ **2.** Type of symbiosis in which one individual benefits while the other individual is harmed

_______________ **3.** Occurs when one organism captures and eats another organism

_______________ **4.** Type of symbiosis in which both individuals benefit

_______________ **5.** Occurs when two organisms fight for the same limited organisms

_______________ **6.** Type of symbiosis in which one individual benefits while the other individual neither benefits nor is harmed

carrying capacity	immigration	population crash
emigration	limiting factor	

_______________ **7.** The movement of individuals *out* of a population into another population

_______________ **8.** The maximum number of individuals of a certain species that an environment can normally support over a long period of time

_______________ **9.** The movement of individuals *into* a population from another population

_______________ **10.** A dramatic decline in the size of a population over a short period of time

_______________ **11.** A factor that controls the size of a population

Name Period Date

C. Vector Vocabulary Define the words in the boxes. On each arrow, write a phrase that describes how the words in the boxes are related to each other.

SYMBIOTIC

1. ______________________________

____

____

is a type of — is a type of — is a type of

MUTUALISM

2. ______________________________

____

____

COMMENSALISM

3. ______________________________

____

____

PARASITISM

4. ______________________________

____

____

EXAMPLE

5. ______________________________

____

____

____

EXAMPLE

6. ______________________________

____

____

____

EXAMPLE

7. ______________________________

____

____

____

Name Period Date

D. Secret Message Fill in the blanks with the vocabulary word that best fits. When complete, write the boxed letters in order in the blanks at the bottom of the page.

1. All of the abiotic and biotic factors in the area where a species lives

__ [ ] __ __ __ __ __

2. A factor that has the greatest effect in keeping down the size of a population

[ ] __ __ __ __ __ __ __ __

3. The process by which one organism captures and feeds upon another organism

__ __ __ [ ] __ __ __ __ __

4. A type of species that is the first to live in a previously uninhabited area

__ __ __ [ ] __ __ __

5. A type of population growth in which a period of slow growth is followed by a short period of exponential growth before leveling off at a stable size

[ ] __ __ __ __ __ __ __ __

6. Occurs when two individuals compete for the same resources

__ __ __ __ [ ] __ __ __ __ __ __

7. A close relationship between two or more different species that live in close contact with one another

__ __ __ __ __ [ ] __ __ __

8. A symbiotic relationship in which one organism is helped and the other is hurt

[ ] __ __ __ __ __ __ __ __ __

9. The movement of individuals into a population from a different population

__ __ __ __ __ __ __ __ [ ] __

10. A symbiotic relationship in which both organisms benefit

__ __ __ __ __ [ ] __ __

11. A type of succession in which an ecosystem damaged by fire is reestablished

__ __ __ __ __ [ ] __ __ __

Fill in the blanks with the boxed letters from above to name the famous ecologist:

__ __ __ __ __ __ __ __ __ __ __

CHAPTER 14
Interactions in Ecosystems

Name Period Date

SECTION **15.1** | LIFE IN THE EARTH SYSTEM

Study Guide

KEY CONCEPT

The biosphere is one of Earth's four interconnected systems.

VOCABULARY		
biosphere	hydrosphere	geosphere
biota	atmosphere	

MAIN IDEA: The biosphere is the portion of Earth that is inhabited by life.
Write a description of each Earth system in the table below.

Earth System	Description
1. biosphere	
2. hydrosphere	
3. atmosphere	
4. geosphere	

5. What is the connection between the biota and the biosphere?

6. Use an example to explain how the four Earth systems are connected.

CHAPTER 15 The Biosphere

7. Fill in the following diagram with the correct term (biosphere, biota, hydrosphere, atmosphere, geosphere).

MAIN IDEA: Biotic and abiotic factors interact in the biosphere.

8. In your own words, describe the Gaia hypothesis.

____

____

____

Vocabulary Check

Choose the word from the box below that best matches up with each Earth system.

air	water	earth	life

9. Atmosphere ________________

10. Biosphere ________________

11. Geosphere ________________

12. Hydrosphere ________________

Name Period Date

SECTION 15.1 | LIFE IN THE EARTH SYSTEM

Power Notes

Earth System	Description

Scientists who contributed to the Gaia hypothesis:

- ____________________________
- ____________________________

Gaia hypothesis summary:

Name Period Date

SECTION **15.1** | LIFE IN THE EARTH SYSTEM

Reinforcement

KEY CONCEPT The biosphere is one of Earth's four interconnected systems.

The **biosphere** is the part of Earth where life exists. Within the biosphere is a collection of living things called the **biota**. The biosphere is connected to three other Earth systems:

- the **hydrosphere**, which includes all of Earth's water, ice, and water vapor
- the **atmosphere**, which includes the air blanketing the surface of Earth
- the **geosphere**, which includes all of the features of Earth's surface and everything below the surface of Earth

Biotic and abiotic factors interact in the biosphere. A change in one Earth system can affect the others. James Lovelock proposed the Gaia hypothesis to explain how biotic and abiotic factors interact in the biosphere. In this hypothesis, the Earth is considered to be a living organism in which the atmosphere, geosphere, and hydrosphere are cooperating systems that yield a biosphere capable of supporting life. The Gaia hypothesis recognizes the complex connections and feedback loops between the biotic and abiotic components of Earth.

1. List the four Earth systems.

____

____

2. What is the connection between the biota and the biosphere?

____

____

____

3. What is the Gaia hypothesis?

____

____

____

Name Period Date

SECTION
15.2 | CLIMATE
Study Guide

KEY CONCEPT
Climate is a key abiotic factor that affects the biosphere.

VOCABULARY	
climate	microclimate

MAIN IDEA: Climate is the prevailing weather of a region.

1. What is the difference between an area's weather and climate?

____

____

2. What are four key factors that shape an area's climate?

____

MAIN IDEA: Earth has three climate zones.

3. Name the main reason why the surface of Earth is heated unevenly by the Sun.

____

4. What characteristic of Earth results in different seasons over a period of a year?

____

____

Complete the following chart with the location and characteristics of each climate zone.

Zone	Location	Characteristics
5. polar zone		
6. tropical zone		
7. temperate zone		

CHAPTER 15
The Biosphere

8. What effect does the heating of Earth have on air and water movement?

9. Why do areas closer to bodies of water have different climates than do inland areas?

10. How does the presence of mountains affect an area's climate?

11. What is a rain shadow?

Vocabulary Check

12. What is the difference between a climate and a microclimate?

13. List four characteristics of the climate where you live. Include information on temperature and precipitation.

Name Period Date

SECTION 15.2 | CLIMATE

Power Notes

Climate is:

Microclimate is:

Climate Zones

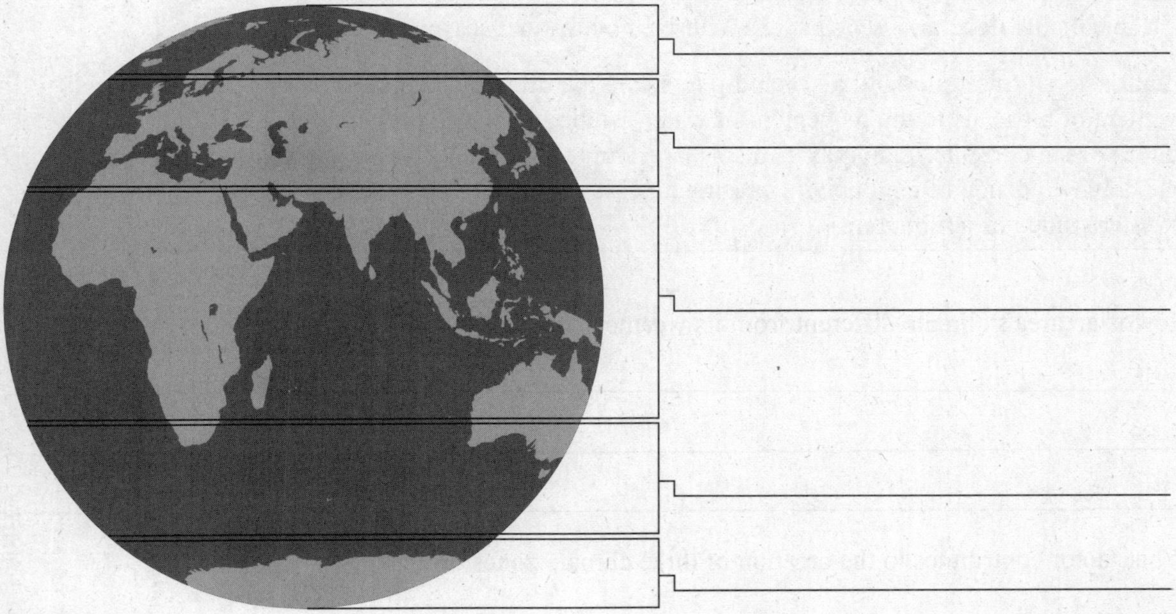

Factors that influence climate:

- ____________________
- ____________________
- ____________________

Name Period Date

SECTION
15.2 CLIMATE
Reinforcement

KEY CONCEPT Climate is a key abiotic factor that affects the biosphere.

While weather changes on a daily, and sometimes hourly, basis, **climate** is defined as the long-term pattern of weather conditions in a region. An area's climate includes factors such as average temperature, average precipitation, and relative humidity. A **microclimate** is the climate of a small specific place within a larger area.

Due to Earth's curved shape, the planet is heated unevenly by the Sun, creating three main climate zones. These zones are determined by the angle in which sunlight hits Earth. The three zones are

- the polar climate zone, located in far northern and far southern reaches of the planet, where the temperature is often below freezing
- the tropical climate zone, located at the equator, which is characterized by warm, moist conditions
- the temperate climate zone, located in the wide area that lies between the polar and tropical climate zones, which is characterized by distinct seasons of equal length

Sunlight also warms water and air, helping to shape the different climate zones. Movement of air leads to the movement of water, which, along with other factors, produces ocean currents. Landmasses also shape climates. A rain shadow is produced on the downwind side of a mountain, causing an eastern slope to be much drier than the western slope of a mountain.

1. How is an area's climate different from its weather?

____

____

____

2. What factor contributes to the creation of three climate zones on Earth?

____

____

3. What are the three main climate zones?

____

____

Name Period Date

SECTION
15.3 | BIOMES
Study Guide

KEY CONCEPT
Biomes are land-based, global communities of organisms.

VOCABULARY	
canopy	coniferous
grassland	taiga
desert	tundra
deciduous	chaparral

MAIN IDEA: Earth has six major biomes.
Fill in the chart with details about the six major biomes found on Earth.

Biome	Description
1. tropical rain forest	
2. grassland	
3. desert	
4. temperate	
5. taiga	
6. tundra	

7. What is the difference between tropical and temperate grasslands?

8. What are the four different types of deserts?

9. How does precipitation differ in a temperate deciduous forest and a temperate rain forest?

____

____

10. Why do few plants grow in the tundra?

____

____

11. Describe the main characteristics of chaparral.

____

____

MAIN IDEA: Polar ice caps and mountains are not considered biomes.

12. Why aren't polar ice caps and mountains considered biomes?

____

____

13. Where are the polar ice caps located?

____

____

14. What is a mountain life zone?

____

____

Vocabulary Check

15. I lose my leaves in the autumn. I am a ____________________.

16. I retain my needles all year long. I am a ____________________.

17. I am the uppermost branches of a tree. I am called the ____________________.

Name Period Date

SECTION
15.3

BIOMES
Power Notes

Tropical	Description
	• •
Grassland	• •
	• •
Desert	• •
Temperate	• •
	• •
Taiga	• •
Tundra	• •

Name Period Date

SECTION
15.3 | BIOMES
Reinforcement

KEY CONCEPT Biomes are land-based, global communities of organisms.

A biome is a major community of organisms, usually characterized by the climate conditions and plant communities that live there. Earth has six major biomes. These broad biome types can be subdivided into even more specific zones. Climate conditions of the Earth's biomes include

- Tropical rain forest—warm temperatures and abundant rainfall occur all year long
- Tropical **grassland**—warm temperature throughout the year, with specific rainy and dry seasons
- Temperate grassland—dry and warm during the summer, most precipitation falls as snow during the winter
- **Desert**—very dry climate all year long
- Temperate **deciduous** forest—hot temperatures in the summer and cold temperatures in the winter; precipitation occurs evenly across the year
- Temperate rain forest—one long wet season and a relatively dry summer
- **Taiga**—long, cold winters and short, warm and humid summers
- **Tundra**—subzero temperatures during the long winter, and little precipitation falls across the year

Chaparral is a minor biome that is characterized by hot, dry summers and cool, moist winters. Most plants found in this biome are small-leaved evergreen shrubs.

Polar caps and mountains are not considered biomes. Polar caps, which are found at the poles at the top and bottom of Earth are ice-covered areas that have no soil and no specific plant community. Mountains are not considered biomes because climate conditions change on a mountain as elevation increases.

1. What is a biome?

____

____

2. List eight biomes that occur on Earth.

____

____

____

3. Why aren't ice caps or mountains considered biomes?

____

____

Name Period Date

SECTION 15.4 | MARINE ECOSYSTEMS

Study Guide

KEY CONCEPT

Marine ecosystems are global.

VOCABULARY		
intertidal zone	abyssal zone	phytoplankton
neritic zone	plankton	coral reef
bathyal zone	zooplankton	kelp forest

MAIN IDEA: The ocean can be divided into zones.

Complete the following table with information about ocean zones.

Zone	Depth	Description
1. intertidal		
2. neritic		
3. bathyal		
4. abyssal		

5. What zone has the most biomass? What type of organism makes up most of this biomass?

____

____

____

6. Why are phytoplankton critical to life on Earth?

____

____

____

CHAPTER 15 The Biosphere

MAIN IDEA: Coastal waters contain unique habitats.

7. Complete the following Y-diagram to outline the similarities and differences between a coral reef and a kelp forest.

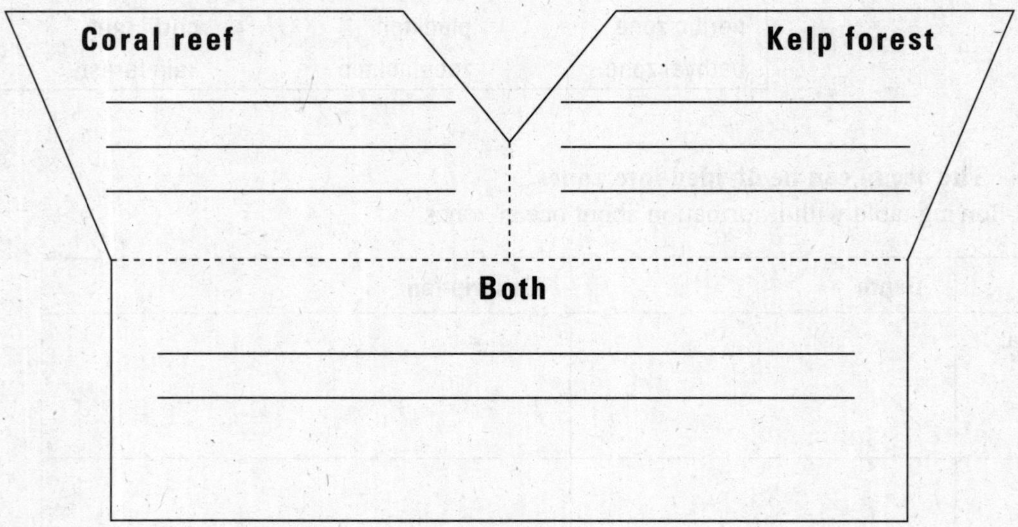

8. What is a coral reef made from?

____

____

9. Why are coral reefs considered delicate?

____

____

Vocabulary Check

10. I am a photosynthetic plankton. What am I? ____________________

11. I am an animal plankton. What am I? ____________________

Name Period Date

SECTION
15.4

MARINE ECOSYSTEMS

Power Notes

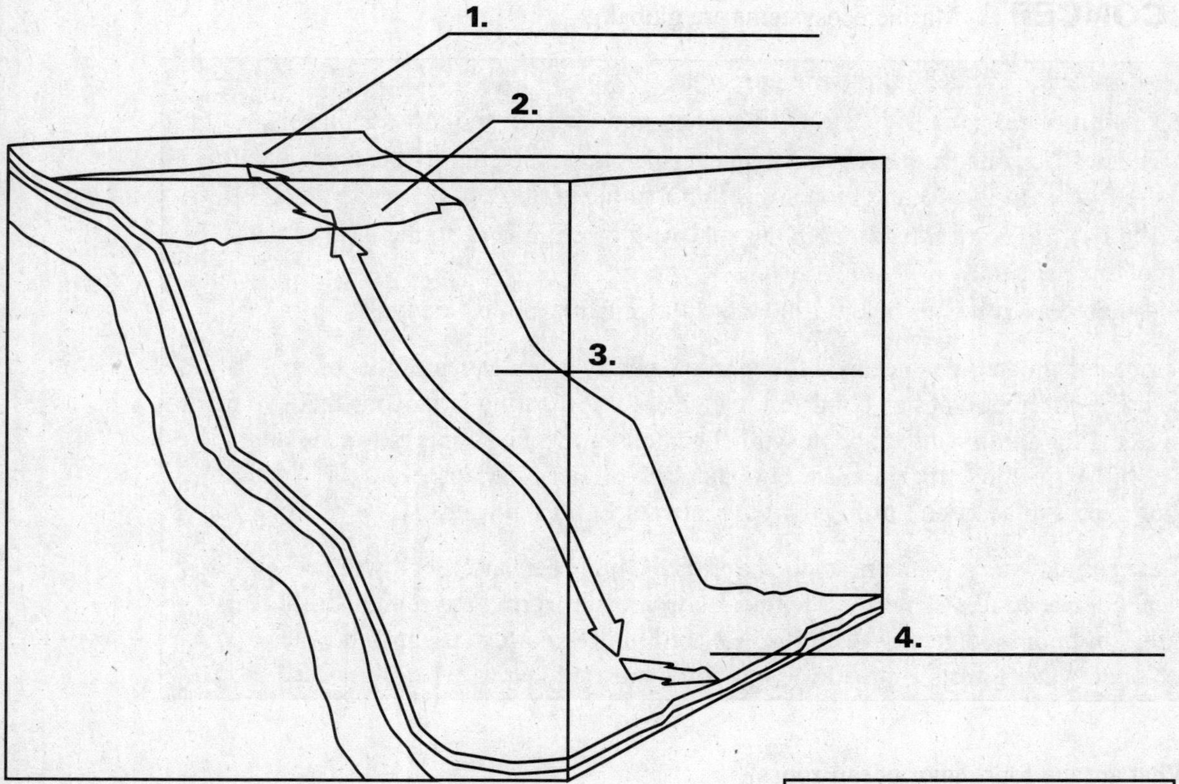

Two unique coastal habitats:

-
-

Description of Ocean Zones

1.	2.
3.	4.

CHAPTER 15
The Biosphere

Name Period Date

SECTION **15.4** | MARINE ECOSYSTEMS

Reinforcement

KEY CONCEPT Marine ecosystems are global.

The ocean can be divided into four major zones:

- The **intertidal zone** is the strip of land between the high and low tide lines.
- The **neritic zone** extends from the intertidal zone to the edge of the continental shelf; most of the ocean's biomass is found in the neritic zone.
- The **bathyal zone** extends from the edge of the neritic zone to the base of the continental shelf.
- The **abyssal zone** lies below 2000 meters and is in complete darkness.

Most of the biomass in an ocean is found in the neritic zone. Much of this biomass is made up of different types of **plankton**, which are free-floating organisms that live in the water. **Phytoplankton** are photosynthetic plankton, while **zooplankton** are animal plankton. Marine phytoplankton are critical to life on Earth because they carry out the bulk of photosynthesis on Earth, providing most of Earth's oxygen.

Shallow coastal waters contain unique habitats. **Coral reefs** are found within the tropical climate zone, where water temperatures remain warm year-round. Coral reefs are areas of high biodiversity. **Kelp forests** are found in cold, nutrient-rich waters. These underwater forests are made up of communities of kelp, a type of seaweed.

1. What are the four major ocean zones?

____

2. Which oceanic zone contains the most biomass?

____

3. In terms of their source of energy, what is the difference between phytoplankton and zooplankton?

____

____

____

4. Why wouldn't you find a kelp forest near a coral reef?

____

____

____

Name Period Date

SECTION **15.5** | ESTUARIES AND FRESHWATER ECOSYSTEMS

Study Guide

KEY CONCEPT

Freshwater ecosystems include estuaries as well as flowing and standing water.

VOCABULARY		
estuary	littoral zone	benthic zone
watershed	limnetic zone	

MAIN IDEA: Estuaries are dynamic environments where rivers flow into the ocean.

1. What is an estuary?

2. What is the distinctive feature of an estuary?

3. Describe why estuaries are considered to be highly productive ecosystems.

4. Why are estuaries sometimes called the "nurseries of the sea"?

5. What adaptations are necessary for organisms that live in an estuary?

6. What impact does the removal of an estuary have on surrounding areas?

CHAPTER 15
The Biosphere

MAIN IDEA: Freshwater ecosystems include moving and standing water.

7. What are the characteristics of a wetland?

8. What is an important function of wetlands with regard to the water supply?

MAIN IDEA: Ponds and lakes share common features.

9. Complete the following chart with details about the different zones found in a pond or lake.

Zone	Location	Description
littoral zone		
limnetic zone		
benthic zone		

Vocabulary Check

10. What is a watershed?

11. The term *estuary* comes from the Latin word *aestus*, which means "tide." How does this meaning relate to the definition of estuary?

Name Period Date

SECTION 15.5

ESTUARIES AND FRESHWATER ECOSYSTEMS

Power Notes

Definition	Description

Estuary

Other Facts	Threats

Lake Zones

Zone	Description

CHAPTER 15
The Biosphere

Name Period Date

SECTION **15.5** | ESTUARIES AND FRESHWATER ECOSYSTEMS

Reinforcement

KEY CONCEPT Freshwater ecosystems include estuaries as well as flowing and standing water.

An **estuary** is a partially enclosed body of water formed where a river flows into an ocean. The distinctive feature of an estuary is the mixture of fresh water from a river with salt water from the ocean. Because of the high amount of nutrients brought into an estuary from the river and the ocean, estuaries are extremely productive ecosystems. Estuaries also provide a sanctuary for animals to reproduce and re-fuel. Estuary ecosystems also act as a buffer between the ocean and coastal lands and help to prevent flooding that results from storms such as hurricanes.

Rivers and streams are flowing bodies of water that serve as pathways through a number of different ecosystems. A **watershed** is a region of land that drains into a river, river system, or other body of water. Wetlands are freshwater ecosystems characterized by the presence of standing water, or water that flows very slowly.

Like oceans, freshwater lakes and ponds can also be divided into separate zones:

- The **littoral zone** is located between the high and low water marks along the shoreline, and its warm and well-lit waters are the home to a number of plants and animals.
- The **limnetic zone** refers to the open water located farther out from shore, and is characterized by an abundance of plankton communities, which support fish populations.
- The **benthic zone** is the lake or pond bottom, where less sunlight reaches, and is inhabited by decomposers such as bacteria.

1. What is an estuary?

____

____

2. What is the distinctive feature of an estuary?

____

3. What is a wetland?

____

____

4. What are the three zones that make up a lake or pond?

____

Name Period Date

CHAPTER 15 | CONSTRUCTING COMBINATION GRAPHS: CLIMATOGRAM

Data Analysis Practice

Climatograms show average climate data for a specific location or biome collected over a period of time.

Climate Data for Sydney, Australia, 1858–2004

Month	Temperature (°C)	Precipitation (mm)
January	22.2	103.3
February	22.2	117.4
March	21.1	131.2
April	18.6	127.2
May	15.4	123.3
June	13.1	128.1
July	12.1	98.1
August	13.3	81.5
September	15.5	68.7
October	17.8	76.9
November	19.6	83.1
December	21.3	78.1

Source: Australian Government Bureau of Meteorology

1. **Construct** Use the information from the table to construct your own climatogram in the space below.

Name Period Date

2. Identify During which three months did the least amount of precipitation fall?

3. Conclude What pattern exists in the data?

CHAPTER 15
The Biosphere

CHAPTER 15 | CLIMATE CHANGE CONTROVERSY

Pre-AP Activity

In recent years, the controversial issues of global warming and climate change have been widely covered by the mass media. Thousands of newspaper, magazine, and journal articles have been written. Local and national news stations have broadcast numerous segments and specials addressing the issue. Hundreds of Web sites are dedicated to providing information regarding global warming and climate change. Mainstream movies such as "The Day After Tomorrow" and novels such as Michael Crichton's *State of Fear* have focused on these issues. When you consider all of the different sources of information on these issues, how do you know what to think?

GLOBAL WARMING AND CLIMATE CHANGE

Global warming is an increase in the average temperature of the Earth's atmosphere, which can result in climatic change. Climate change refers to any significant change in the measures of climate (temperature, precipitation, or wind) over an extended period of time (decades or longer). Data collected by NOAA and NASA show that the average temperature of the Earth's surface has increased 1.2 to 1.4 degrees (F) since 1900. In addition, changes in precipitation patterns, snow and ice cover, and sea levels indicate that climate change is taking place.

According to the EPA, climate change can be caused by natural factors such as changes in the Sun's intensity or the Earth's orbit, or volcanic eruptions; natural processes within the climate system such as changes in ocean currents and circulation; and human activities that change the composition of the Earth's atmosphere and land surface. The debate about global warming and climate change centers on these factors. Some people think that climate change is caused by natural processes and is cyclical. Others think that human activities (for example, burning fossil fuels) have increased the levels of greenhouses gases such as carbon dioxide, causing atmospheric temperatures to increase. The Intergovernmental Panel on Climate Change (IPCC), formed in 1988 by the United Nations Environment Programme and the World Meteorological Organization, has stated that "There is new and stronger evidence that most of the warming observed over the last 50 years is attributable to human activities."

MIXED MESSAGES

Journalists working for the mainstream media aim to be fair, balanced, and unbiased in their reporting. This generally means that when an article is written regarding an issue, the most common position relating to the issue is identified and then alternative positions are discussed if they exist and are considered credible. A recent study has found that the U.S. media coverage of global warming using "balanced" reporting can lead to an informational bias. By presenting competing points of view on a scientific issue, both views appear to have equal scientific support and value when one may in fact be supported much less. This type of reporting allows skeptics to challenge and downplay scientific data and understanding, making it difficult for the public to accurately analyze the information.

In this study, published in the July 2004 issue of the journal *Global Environmental Change,* 636 randomly selected news articles relating to human contributions to global warming were examined. All of the selected articles had been published in either the *New York Times, Washington Post, Los Angeles Times,* or *Wall Street Journal* between the years 1988 and 2002. The study found that 53% of the examined articles gave approximately equal attention to the opposing views that global warming is the result of natural fluctuations and

Name Period Date

that humans are merely a contributing factor. It was also determined that 35% presented both sides of the issue, but focused on human contributions. Six percent of the articles focused explicitly on the predominant scientific view that humans are contributing to global warming, while another 6% focused on whether human–caused global warming even exists. In general, the study found that the news coverage deviated considerably from the IPCC's consensus regarding human involvement in global warming.

Answer the following questions on a separate piece of paper.

1. Read the following statement and then explain how it may present an informational bias.

> The ability to study climatic patterns has been critical to the debate over the phenomenon called "global warming." Some scientist believe—and some ice core studies seem to indicate—that humanity's production of carbon dioxide is leading to a potentially dangerous overheating of the planet. But skeptics contend there is no evidence the warming exceeds the climate's natural variation.
>
> *Los Angeles Times,* December 2, 2002.

2. There is substantial scientific data and a consensus within the scientific community that human activities are changing the composition of the Earth's atmosphere and contributing to global warming. From your perspective, explain how informational bias regarding global warming and climate change may affect the general public's understanding of these issues and influence their behavior.

3. In your opinion, do you think that both sides of a scientific issue should be reported on with equal weight when consensus has formed to support just one position? Should journalists always present two or more sides to a story if there only appears to be one? Explain.

Name Period Date

CHAPTER 15 | GLOBAL WARMING AND METHANE EMISSION

Pre-AP Activity

You have learned in Chapter 15 that the tundra biome is a vast barren region located in the far northern latitudes. It comprises one of Earth's harshest habitats: a region of low biotic diversity, meager precipitation, and below-freezing winters that last up to 10 months a year. But while tundra harbors a relatively low diversity of living things, it is rich in organic matter, much of which is locked inside the upper layers of the frozen earth.

TUNDRA CHARACTERISTICS AND ECOLOGY

The ground of the tundra region is a mix of soil, rocks, and ice that, depending on its water content, may resemble frozen mud or cold, dry earth. The uppermost layer of ground is called the active layer. It ranges in thickness from several inches to five feet or more, depending on factors such as local climate and vegetation. In many areas, the active layer may contain or be covered with thick accumulations of peat—partially decayed plant matter rich in carbon. The active layer contains a large proportion of ice. During the short Arctic summer, sections of this may thaw, producing small lakes and bogs that refreeze when temperatures drop with the onset of winter. Below the active layer lies permafrost—ground that remains frozen year round. Permafrost generally resembles chunks of rock and soil cemented together with ice. It extends down at least 300 meters (1000 feet), but may be deeper in some areas.

Tundra winters last 10 months and are extremely harsh—temperatures average around –34°C but may drop below –50°C. Winds can reach speeds of 160 km per hour (100 miles per hour). Days are extremely short; the Sun is completely below the horizon for roughly 10 weeks, keeping the region in total and frozen darkness. Summers last between one and two months; but though the season is short, its days are long, with close to 24 hours of sunlight each day. Summer temperatures average around 3–12°C. Precipitation is scarce, however; total annual precipitation, including winter snowfall, averages 15–25 cm.

The long periods of sunlight in the summer allow for almost continuous photosynthesis. Lichens are abundant. Along with tundra moss, they are important producers. Some tundra communities contain sedges, grasses, and small shrubs. Reindeer, musk oxen, snowshoe hares, lemmings, lynx, and snowy owls are adapted for life here, as are polar bears in coastal areas. Many birds migrate to the tundra in summer and leave in winter. Most decomposition is carried out by bacteria and fungi.

METHANE

Methane (CH_4) is a carbon-based gas produced by both natural and human-related processes. It is an end product of digestion and decomposition. Natural sources of methane include wetlands, oceans, swamps, soils, and permafrost. These sources collectively account for only 40 percent of global methane emissions; human-related activities are responsible for the other 60 percent. Fossil fuel production, rice cultivation, waste management, and livestock are among the most critical sources.

Like carbon dioxide, methane is a greenhouse gas—in the atmosphere, it prevents infrared radiation emitted from Earth's surface from escaping the atmosphere. Without this effect, Earth would be too cold to live on. However, the steady increase in greenhouse gas concentration over the past two centuries has produced global warming, an unprecedented rise in global air and surface temperatures. This, in turn, can cause climate change.

Name Period Date

Methane is far less abundant in the atmosphere compared to carbon dioxide, but it packs a greater punch: a methane molecule is more than 20 times more powerful than carbon dioxide in its ability to trap heat. Analysis of ice cores drawn from glaciers shows that methane is more abundant in Earth's atmosphere today than at any point in the last 400,000 years.

TUNDRA METHANE AND GLOBAL WARMING

The vast Arctic tundra harbors a significant store of carbon compounds. Over the past tens of thousands of years, these compounds were locked inside the permafrost and ice covering the area. The extensive peat bogs are a ready source of carbon; furthermore, when these freeze in winter, bog gases are trapped inside bubbles. Over the past several decades, increased global temperatures have affected the natural thaw/freeze cycles of the tundra. Historically, the increased sunlight and warmer temperatures of summer would cause relatively small lakes and bogs to form as the active layer ice melted; the rapid onset of winter temperatures would cause these to refreeze. However, the thaw-freeze cycle is increasingly skewed: each spring a larger number of lakes and bogs form or increase in size, and each winter, the high levels of gas prevent these areas from refreezing completely. This shift in the natural cycle has produced a serious problem: the more lakes form, the greater the quantity of methane released when the lakes melt.

A study undertaken by a team of Russian and American scientists revealed that lakes in Siberia that thaw every summer are releasing five times more methane than had been estimated. In one study, scientists reported that methane was literally bubbling in the active layer at such a rapid rate that it prevented the surface from refreezing in winter. More than one million square kilometers of tundra have started thawing for the first time in roughly 11,000 years. If the present circumstances continue, methane emissions will rise exponentially, increasing the concentration of atmospheric greenhouse gases, which in turn will raise global temperatures. The increased thaw area in the Arctic poses a further danger: thawing lakes are darker than the surrounding tundra, so their darker color attracts and absorbs more heat, thereby increasing the thaw rate and subsequent emission of methane.

FEEDBACK LOOPS

A feedback loop is a pathway of two or more steps in which the effects of one factor feeds back into the pathway, increasing or decreasing the effects of other factors. In a positive feedback loop, the effect of one factor increases the next factor, which increases the next factor, and so on.

The linked pathway is modeled as a loop because at some point, a link will feed back to the initial factor. The cycle becomes continuous, with no true beginning or end. In some circumstances, positive feedback loops can be forced to end if one or more links can be broken. However, if the effect of one or more stages passes a tipping point, the cycle will "run away"—that is, there will be no possibility of stopping it from continuing on.

DRAW A FEEDBACK LOOP

1. On a separate sheet of paper, model the relationship between tundra methane emissions and climate change by drawing a detailed positive feedback loop. Your loop should feature the factors described in this activity's introductory text.

Name Period Date

CHAPTER 15

THE BIOSPHERE

Vocabulary Practice

biosphere	grassland	neritic zone	estuary
biota	desert	bathyal zone	watershed
hydrosphere	deciduous	abyssal zone	littoral zone
atmosphere	coniferous	plankton	limnetic zone
geosphere	taiga	zooplankton	benthic zone
climate	tundra	phytoplankton	
microclimate	chaparral	coral reef	
canopy	intertidal zone	kelp forest	

A. Word Origins Circle the Greek and Latin word parts in each vocabulary term. Then use the Greek and Latin meanings to construct a very basic definition of the vocabulary word.

abyss-	=	bottomless	bio-	=	life	klima-	=	climate
atmos-	=	vapor	decidere	=	to fall off	limne	=	lake
bathy-	=	deep	geo-	=	earth	litor-	=	shore
benthos	=	bottom	hydro-	=	water	micros	=	small

WORD	DEFINITION
1. atmosphere	
2. hydrosphere	
3. geosphere	
4. biosphere	
5. benthic zone	
6. limnetic zone	
7. littoral zone	
8. bathyal zone	
9. abyssal zone	
10. deciduous	
11. microclimate	

Name Period Date

VOCABULARY PRACTICE, CONTINUED

B. Who Am I? Choose among these terms to answer the riddles below:

atmosphere	desert	kelp forest
canopy	deciduous	hydrosphere
chaparral	estuary	taiga
coniferous	geosphere	tundra
coral reef	grassland	watershed

1. I am a type of tree that drops its leaves during autumn: ____________

2. I am also called the boreal forest; I have long and cold winters: ____________

3. I am an Earth system made up of all of Earth's water, ice, and water vapor: ____________

4. I am also called the Mediterranean shrubland; I have hot, dry summers and cool, moist winters: ____________

5. I am an area where the primary plant life is grass: ____________

6. I am a unique coastal habitat found in cold, nutrient-rich waters; I am made up of communities of seaweed: ____________

7. I am a partially enclosed body of water formed where a river flows into an ocean: ____________

8. I am an Earth system made up of all the features on Earth's surface, including the continents, rocks, and sea floor; and everything below Earth's surface: ____________

9. I am a type of tree that retains its needles all year long: ____________

10. I am a biome located in far northern latitudes; I am covered by a layer of permafrost: ____________

11. I am the uppermost branches of trees: ____________

12. I am an Earth system made up of all the air that covers Earth: ____________

13. I am a region that drains into a river, river system, or another body of water: ____________

14. I am a biome with a very dry, arid climate: ____________

15. I am a unique coastal habitat found in the tropical zone; I support large communities of tropical fish: ____________

C. Put It in a Box For each vocabulary term, fill in the box with words that will help you to remember its meaning.

DECIDUOUS	TUNDRA	CHAPARRAL
1. ______	2. ______	3. ______

CONIFEROUS	TAIGA	ESTUARY
4. ______	5. ______	6. ______

DESERT	BIOSPHERE	CLIMATE
7. ______	8. ______	9. ______

D. Categorize Words List the vocabulary words that belong in each category.

abyssal zone	desert	littoral zone
atmosphere	geosphere	neritic zone
bathyal zone	hydrosphere	phytoplankton
benthic zone	intertidal zone	taiga
biosphere	kelp forest	tundra
coral reef	limnetic zone	zooplankton

Ocean Zones	Earth Systems	Lake Zones
a.	a.	a.
b.	b.	b.
c.	c.	c.
d.	d.	
Coastal Habitats	**Types of Plankton**	**Biomes**
a.	a.	a.
b.	b.	b.
		c.

E. Find the Odd Word Place a check mark next to the word that does not belong and explain why.

1. _____ tundra Explanation ______________________________

   _____ coniferous ______________________________

   _____ taiga

2. _____ coral reef Explanation ______________________________

   _____ desert ______________________________

   _____ kelp forest

Name Period Date

SECTION **16.1** HUMAN POPULATION GROWTH AND NATURAL RESOURCES

Study Guide

KEY CONCEPT

As the human population grows, the demand for Earth's resources increases.

VOCABULARY
nonrenewable resource
renewable resource
ecological footprint

MAIN IDEA: Earth's human population continues to grow.

1. Approximately how big is Earth's population now?

____

____

2. Name and give examples of two technologies that have influenced human population growth since 1700.

____

____

MAIN IDEA: The growing human population exerts pressure on Earth's natural resources.

Determine whether the following resources are renewable or nonrenewable. Explain your answer.

3. sun ____
4. oil ____
5. trees ____
6. water ____
7. wind ____
8. corn ____
9. beef ____
10. coal ____

Name Period Date

STUDY GUIDE, CONTINUED

MAIN IDEA: Effective management of Earth's resources will help meet the needs of the future.

11. The inhabitants of Easter Island made many mistakes in their resource use. Name one resource that was misused and describe two ways that they could have used the resource more effectively.

____

12. What is an ecological footprint?

____

13. List the four factors that determine your ecological footprint.

____

Vocabulary Check

14. What is the difference between a renewable and a nonrenewable resource?

____

Be Creative

Create a poster that illustrates why it is important to conserve natural resources.

CHAPTER 16
Human Impact on Ecosystems

Name Period Date

SECTION **16.1** | HUMAN POPULATION GROWTH AND NATURAL RESOURCES

Power Notes

World Population

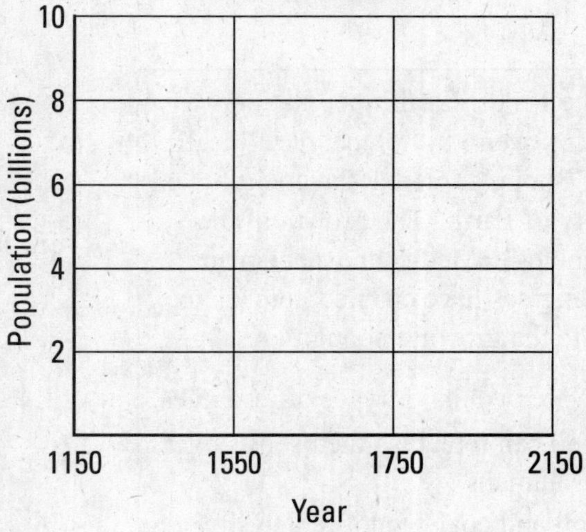

Two technological advancements that have contributed to population growth:

-
-

Types of Resources

Resource Type	Description

Ecological Footprint

Definition:

-

Size depends on:

-

Name Period Date

SECTION **16.1** HUMAN POPULATION GROWTH AND NATURAL RESOURCES

Reinforcement

KEY CONCEPT As the human population grows, the demand for Earth's resources increases.

The human population of Earth continues to grow. In the 1700s, Earth's population was around 1 billion people. Today, this number has growth to over 6 billion people. Recall that the carrying capacity of an ecosystem is the size of a population that the environment can sustain. Scientists do not know the carrying capacity of Earth. The growth of the human population is the result of advancements in technology. Medical advancements help to protect humans from disease, and gas-powered engines have enabled humans to do much more work to provide food and transportation to the growing population.

A large population uses a great deal of resources. There are two major types of resources:

- **Renewable resources** such as the sun, wind, and soil can replenish themselves over a short period of time and continue to be useful for humans.
- **Nonrenewable resources** such as the fossil fuels oil and coal cannot replenish themselves and are being used faster than they form.

Managing Earth's renewable and nonrenewable resources is important for the human population. The overuse of renewable resources can turn them into nonrenewable resources and may become a major problem in the future. Every human on Earth has an ecological footprint. An **ecological footprint** is the amount of land that is needed to produce and maintain enough food and water, shelter, energy, and waste for each person. You can more easily think of your ecological footprint as everything in your lives that came from a natural product. A milk carton, your desk, and your home all came from some place where they took up natural space. Minimizing your ecological footprint will help to conserve renewable and nonrenewable resources.

1. What advancements helped Earth's human population to grow so quickly?

____

____

2. What is the difference between a renewable and nonrenewable resource?

____

____

3. What is included in your ecological footprint?

____

____

Name Period Date

SECTION
16.2 | AIR QUALITY
Study Guide

KEY CONCEPT
Fossil fuel emissions affect the biosphere.

VOCABULARY	
pollution	acid rain
smog	greenhouse effect
particulate	global warming

MAIN IDEA: Pollutants accumulate in the air.

1. What is pollution?

____

____

2. What is smog?

____

____

3. What are the major components of smog and how does it form?

____

____

4. What is acid rain?

____

____

5. How does acid rain affect ecosystems?

____

____

Name Period Date

STUDY GUIDE, CONTINUED

MAIN IDEA: Air pollution is changing Earth's biosphere.

Complete the concept map with information about the greenhouse effect.

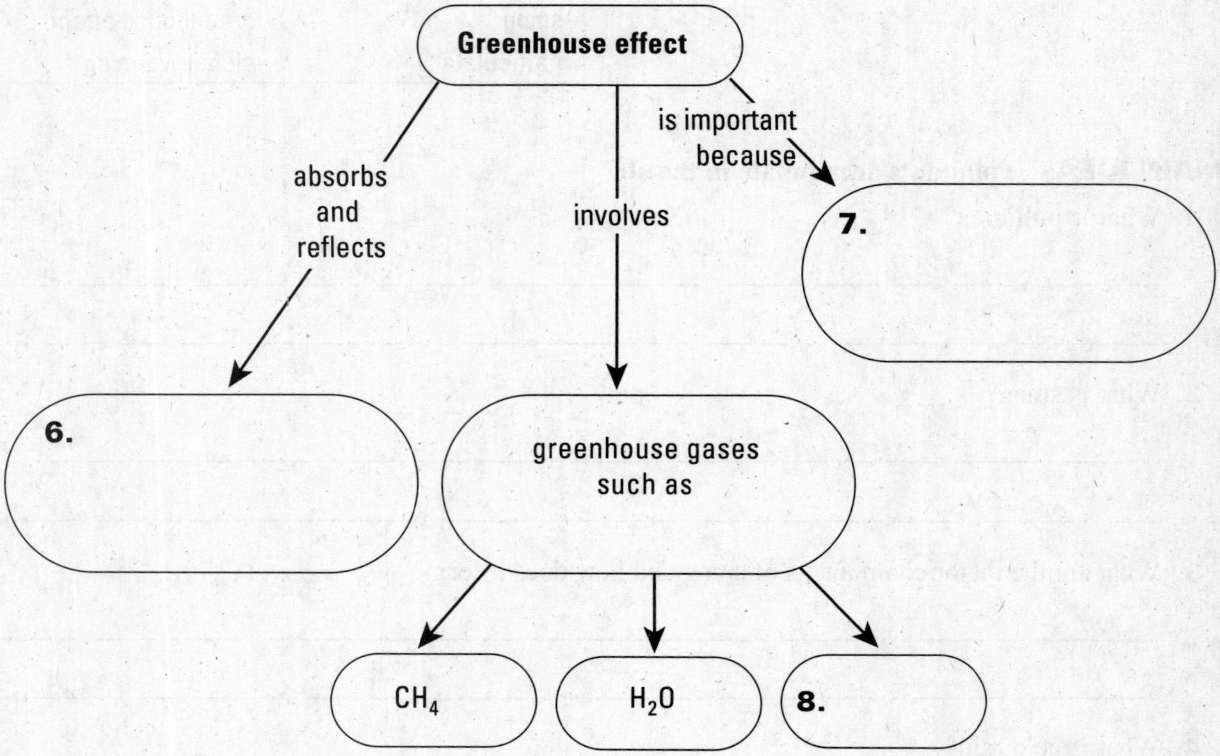

9. What is the greenhouse effect?

10. What is the relationship between the greenhouse effect and global warming?

Vocabulary Check

11. How is a gardener's greenhouse a miniature version of the greenhouse effect?

12. The word *particulate* comes from the Latin word *particula*, which means "a small part." How is this word origin related to the definition of a particulate?

CHAPTER 16
Human Impact on Ecosystems

SECTION
16.2 AIR QUALITY
Power Notes

Greenhouse effect occurs when:

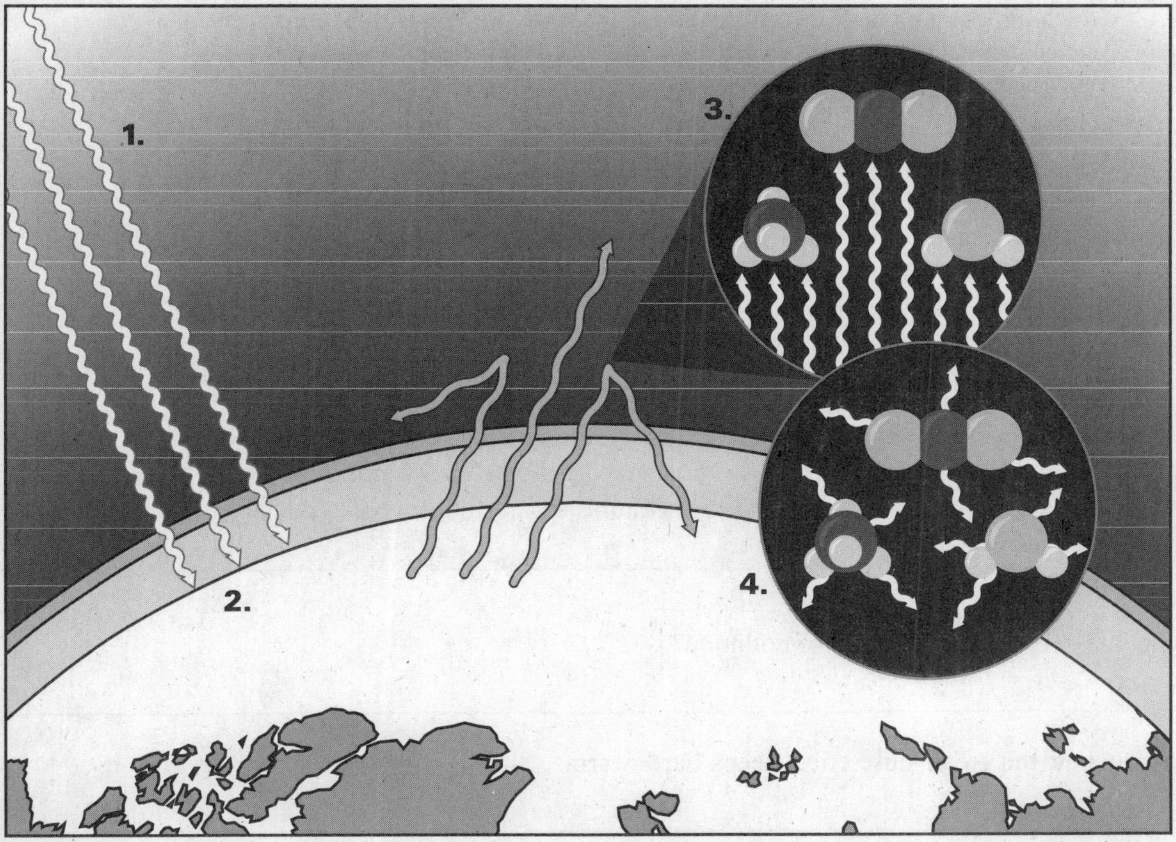

1.

3.

2.

4.

Name Period Date

SECTION **16.2** | AIR QUALITY

Reinforcement

KEY CONCEPT Fossil fuel emissions affect the biosphere.

The air you breathe is filled with molecules of oxygen, hydrogen, and carbon. When the air is polluted, you are breathing in molecules that may be harmful, or toxic, to your health. **Pollution** is the addition of any undesirable factor to the air, water, or soil. There are many types of pollution and it happens all around us.

The most common type of air pollution is smog. **Smog** is a hazy cloud of air pollution caused by the interaction of sunlight with pollutants produced by fossil fuel emissions. Another important type of pollution affects precipitation. **Acid rain** is precipitation produced when pollutants in the water cycle cause rain's pH to drop below normal levels. Acid rain can harm crops, forests, and also lakes and streams. An important phenomenon controls the temperature and climate of Earth. Sunlight heats up the surface of Earth. This heat does not stay on the surface of Earth, rather, it is released as energy, and if Earth's atmosphere was not there to prevent it from leaving, our planet would be very cold. The **greenhouse effect** is a normal process in which greenhouse gases, such as carbon dioxide and methane, absorb some of the energy released by Earth's surface to help keep our planet warm.

The burning of fossil fuels releases carbon dioxide into the atmosphere. Humans burn a lot of fossil fuels. The addition of all this extra carbon dioxide is holding in heat from sunlight for a longer time and the consequence is called global warming. **Global warming** is the trend in increasing global temperatures as a result of increased levels of greenhouse gases.

1. What are two significant types of pollution?

____

2. Explain how the greenhouse effect keeps Earth warm.

____

____

____

____

3. How is global warming related to the greenhouse effect?

____

____

Name Period Date

SECTION
16.3

WATER QUALITY
Study Guide

KEY CONCEPT
Pollution of Earth's freshwater supply threatens habitat and health.

VOCABULARY
indicator species
biomagnification

MAIN IDEA: Water pollution affects ecosystems.

1. List three examples of water pollution.

___

2. Why are indicator species important to scientists?

___

MAIN IDEA: Biomagnification causes accumulation of toxins in the food chain.

3. What is biomagnification?

___

___

4. Illustrate an ecosystem's food chain and describe what will happen to the concentration of pollutants as they move up the food chain.

CHAPTER 16
Human Impact on Ecosystems

Vocabulary Check

5. Use your knowledge of the prefix *bio-* and the term *magnification* to explain the meaning of biomagnification.

Be Creative

6. Design a poster that explains the importance of keeping sources of fresh water free from pollution.

Name Period Date

SECTION
16.3

WATER QUALITY
Power Notes

An indicator species is:

Biomagnification is:

Pollutant Concentration
(Draw an arrow from low to high concentration.)

Trophic Level

____

____

____

____

____

Name Period Date

SECTION
16.3

WATER QUALITY
Reinforcement

KEY CONCEPT Pollution of Earth's freshwater supply threatens habitat and health.

Water is a resource that is very vulnerable to pollution. Runoff from farms and cities collects in streams, lakes, and rivers and can put entire ecosystems and human health at risk.

Scientists use certain species to determine the health of the environment. **Indicator species** are those species that provide a sign, or indication, that there may be a problem with pollution in an ecosystem. In aquatic ecosystems, frogs and fish are major indicator species and may show signs such as tissue damage, or in extreme cases may exhibit mutations such as extra legs or fins.

In some ecosystems, harmful pollutants can affect entire food chains. Even though these pollutants may only be found in small amounts, these small amounts can accumulate in organisms high up in the food chain. Recall that a food chain involves producers, primary consumers, secondary consumers, and tertiary consumers. In aquatic ecosystems, producers take in pollutants and store them in their tissues. A primary consumer eats many producers and all of the pollutants in the producer become a part of the primary consumer. Similarly, this happens to secondary consumers, and tertiary consumers. At the top of a food chain, a tertiary consumer will have accumulated a large concentration of pollutants in its body, and may in fact die or fail to reproduce due to these pollutants. This process is called biomagnification. **Biomagnification** is the process by which pollutants accumulate in larger amounts as they move through the food chain.

1. What is an indicator species?

____

____

2. Explain the process of biomagnification.

____

____

____

____

3. Why are pollutants more harmful to tertiary consumers as opposed to producers?

____

____

Name Period Date

SECTION **16.4** | THREATS TO BIODIVERSITY

Study Guide

KEY CONCEPT

The impact of a growing human population threatens biodiversity.

VOCABULARY
habitat fragmentation
introduced species

MAIN IDEA: Preserving biodiversity is important to the future of the biosphere.

1. What is biodiversity?

____

____

2. Why is it important to preserve biodiversity?

____

____

3. Where are the highest levels of biodiversity on our planet? Explain why this is so.

____

____

MAIN IDEA: Loss of habitat eliminates species.

4. List three ways in which humans cause habitat fragmentation.

____

____

MAIN IDEA: Introduced species can disrupt stable relationships in an ecosystem.

5. What is an introduced species?

____

____

CHAPTER 16 Human Impact on Ecosystems

6. Complete the chart below with examples of introduced species and describe how they are disrupting the ecosystem in which they live.

Species	Impact on Ecosystem
Burmese python (Everglades)	
Kudzu (United States)	
Mice (Australia)	

Vocabulary Check

7. A fragment is defined as "a small part broken off or detached." How does this definition relate to the meaning of habitat fragmentation?

____

____

Be Creative

8. Think of an area where you live that is an example of habitat fragmentation. Design a poster that both illustrates the problem and proposes a solution.

Name Period Date

SECTION
16.4 THREATS TO BIODIVERSITY
Power Notes

Why biodiversity is important:

-
-
-

Threats to Biodiversity

Habitat fragmentation:

Introduced species:

Species	Where Introduced	Problems Caused

CHAPTER 16
Human Impact on Ecosystems

Name Period Date

SECTION **16.4** THREATS TO BIODIVERSITY

Reinforcement

KEY CONCEPT The impact of a growing human population threatens biodiversity.

As humans continue to spread out over the entire globe, they are removing wild habitat to make room for more people. By removing this habitat, human also threaten the survival of many different species of plants, animals, and other organisms. The assortment, or variety, of living things in an ecosystem is called biodiversity. The human alteration of habitats threatens biodiversity.

One way that humans are threatening habitat and biodiversity is by habitat fragmentation. **Habitat fragmentation** occurs when a barrier is formed that prevents individuals from one species from moving throughout their home range. Imagine one day that you are not allowed to go home from school because a river is now in the way. This is a simplistic example of habitat fragmentation.

Another way that humans threaten biodiversity is by the introduction of new and invasive species. An **introduced species** is any species that was brought to an ecosystem as the result of human actions. In many cases, introduced species, or invasive species, can cause great damage to an ecosystem:

- Introduced species may disrupt ecosystem functions by preying on native species that have no defense against them.
- Introduced species may also be better competitors for resources. In some cases they may even push native species to extinction.
- Introduced species may also cause economic damage by harming crops or feeding on food stores.

1. What is biodiversity?

____

____

2. How does habitat fragmentation affect a population?

____

____

____

3. What is an introduced species?

____

____

Name Period Date

SECTION
16.5

CONSERVATION
Study Guide

KEY CONCEPT
Conservation methods can help protect and restore ecosystems.

VOCABULARY
sustainable development
umbrella species

MAIN IDEA: Sustainable development manages resources for present and future generations.

1. How can sustainable development help Earth's human population?

____

____

2. Complete the following chart with two examples of sustainable development and explain how they benefit humans.

Resource	How Is It managed?	Benefits

MAIN IDEA: Conservation practices focus on a few species but benefit entire ecosystems.

3. What is an umbrella species?

____

____

CHAPTER 16
Human Impact on Ecosystems

Complete the concept map with information about the manatee and its role as an umbrella species.

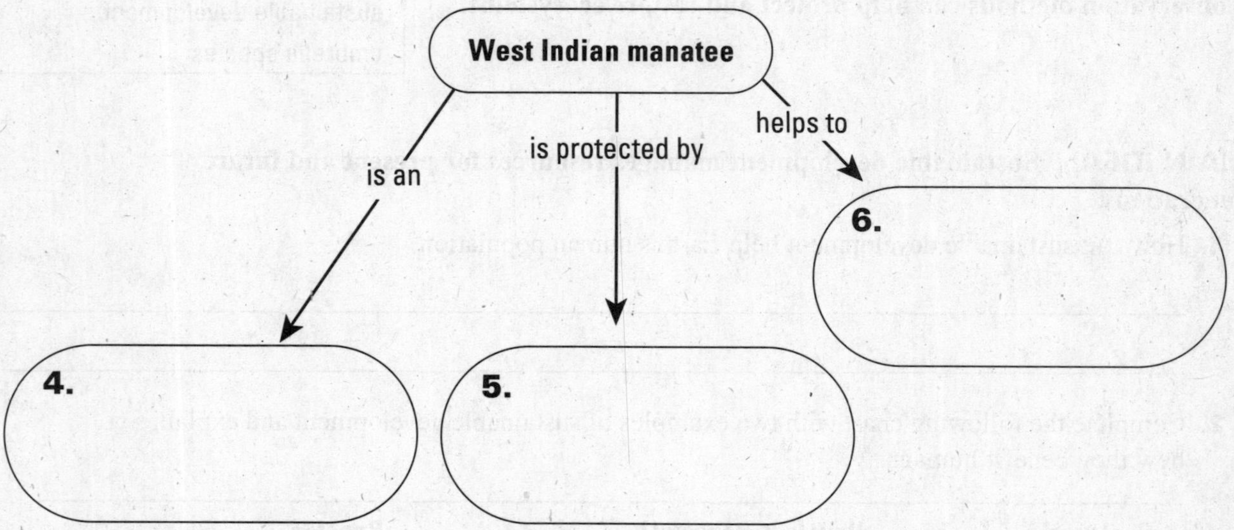

MAIN IDEA: Protecting Earth's resources helps to protect our future.

7. What are three laws that have been developed to help protect natural resources?

8. What can humans do to reduce their impact on Earth's ecosystems?

Vocabulary Check

9. The word *sustain* means "to keep in existence, maintain." How does this meaning relate to the idea of sustainable development?

Name Period Date

SECTION 16.5 | CONSERVATION

Power Notes

Sustainable development is:	Sustainable practices in the fishing industry:
	• • • •

An umbrella species is:

Three important environmental laws:

-
-
-

Ways in which humans can protect the environment:

-
-
-
-

CHAPTER 16
Human Impact on Ecosystems

Name Period Date

SECTION **16.5** CONSERVATION
Reinforcement

KEY CONCEPT Conservation methods can help protect and restore ecosystems.

There are many ways that humans can protect the future of Earth's ecosystems. One way to protect Earth's resources is through sustainable development. **Sustainable development** is a practice in which natural resources are used and managed in a way that meets current needs without hurting future generations. By only using what we need and being careful with the resources we do have, Earth's ecosystems can continue to provide the resources humans need for many years to come.

The preservation of resources can also be accomplished by creating laws to protect environments and species. The Endangered Species Act in the United States is designed to protect species that may be near extinction. In many cases these species also play an important role in their ecosystem. By protecting these **umbrella species** we are also protecting a wide range of other species as well as their habitat. Other laws help to protect important resources:

- The Clean Air Act serves to minimize the amount of pollution that is pumped into our air.
- The Clean Water Act helps to prevent our waterways from being polluted.

Additionally, the establishment of the National Park Service helps set aside areas of wilderness and other lands that are important for our country's heritage. As we move into the future, humans must be aware and take steps to protect the resources Earth provides. Through sustainable development and changing our practices to minimize our use of resources, we can ensure that future generations will be able enjoy planet Earth.

1. What is sustainable development?

____

____

2. How does the protection of an umbrella species benefit an entire ecosystem?

____

____

____

3. What are three laws that have helped to protect Earth's natural resources?

____

____

Name Period Date

CHAPTER 16 | TYPES OF DATA: DISCRETE AND CONTINUOUS

Data Analysis Practice

Data can be discrete or continuous. Discrete data are usually expressed in whole numbers or categories. Continuous data are fractional.

GRAPH 1. AIR QUALITY FOR RIVERSIDE COUNTY, CA 2005

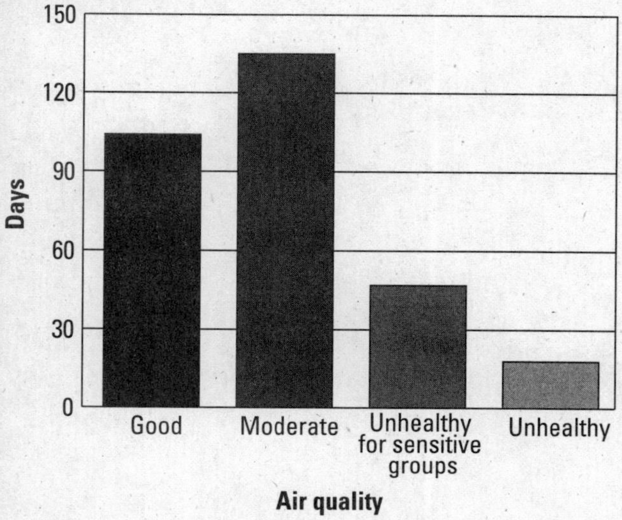

1. Classify Is the dependent variable discrete or continuous? Explain your answer.

____

____

____

____

____

2. Evaluate Suppose the data for air quality was expressed as a percent instead of days. For example, in 2005, 34 percent of the days had good air quality. Would this change the classification of the data as discrete or continuous?

____

____

Name Period Date

CHAPTER 16

BIOMAGNIFICATION OF FLUORINE IN PENGUINS

Pre-AP Activity

In his book *The Crystal Desert: Summers in Antarctica,* biologist David G. Campbell describes how the chemical fluorine is magnified up the Antarctic marine food chain, from krill to penguin:

> Ecologists describe krill as the "keystone" species of the Southern Ocean. They transform diatoms into food eaten by just about every other large predator in the Southern Ocean. But along with being universally appetizing, krill are toxic because they contain high concentrations of the element fluorine, a highly reactive chemical relative of bromine and chlorine (both of which are used to disinfect drinking water and swimming pools). Fluorine is harmless in small quantities; indeed, for humans, ingesting a milligram per day helps prevent tooth cavities. But in quantities greater than ten milligrams per day, fluorine is poisonous, inhibiting enzymes, diminishing growth and fertility, and, because it concentrates in bones, deforming the skeleton. Krill scavenge fluorine from seawater (which contains about one milligram of fluorine per kilogram) and concentrate it in their chitin shells, where levels exceed 3,000 milligrams per kilogram.
>
> Everything that eats krill ingests potentially harmful levels of fluorine. An Adélie penguin, which is about one-tenth the weight of a human, ingests about 240 milligrams of fluorine per day from the krill that it eats. How does it deal with this toxic load? One of the best strategies is simply to rapidly warm the ingested krill with body heat. When the krill die, decomposition causes the fluorine to migrate rapidly from the chitin into the digestible soft tissues; however, the enzymes that release fluorine from the cuticle are denatured at temperatures above 30º C. A penguin's internal body temperature is 38–40º C, so much of the fluorine remains in the indigestible chitin and is excreted in the feces. Most birds, including ducks and chickens, have gastric ceca that enable them to digest cellulose (and its chemical relative chitin). But penguins lack ceca and pass the chitin undigested through their gut. Also, it takes only three to four hours for a krill shell to pass through an Adélie's gut, minimizing the potential for absorption of fluorine. Yet even these adaptations aren't enough, and penguins do absorb high levels of fluorine, which is sequestered in the bones until it can be secreted by the kidneys.

The Crystal Desert: Summers in Antarctica by David G. Campbell.

Campbell goes on to describe how humans are considering how to improve existing krill fisheries and develop new ones to help feed the growing human population, but the high level of fluorine in krill poses a problem. Even when frozen, the fluorine in the chitinous shells of krill can migrate to the meat. This means that even though humans do not eat the shells of krill, they could end up ingesting much of the fluorine that was at one time sequestered in the inedible shell.

Name Period Date

1. How many times greater is the concentration of fluorine in the chitin of krill than in seawater? ____________

2. If a typical Adélie penguin ingests 240 milligrams of fluorine every day from the krill that it eats, how many kilograms of krill must it be eating each day? ____________

3. What are three adaptations that allow the Adélie penguin to minimize the absorption of fluorine from the krill that they eat?

4. In order to prevent fluorine from migrating into the krill meat, what type of processing might need to occur on fishing boats immediately after krill have been caught?

5. The krill that some humans target through commercial fishing are an essential part of the diet of whales, including the endangered blue whale. What adaptation that minimizes the absorption of fluorine might the blue whale have in common with the penguin?

6. In terms of biomagnification of toxins up a food chain, why might the warm body temperature of the blue whale, an animal that can grow to 100 feet, be an inadequate defense against fluorine absorption?

7. Leopard seals are top level predators in the Antarctic marine food chain. They eat penguins, squid, fish, sea birds, and other seals, many of which feed on krill. Young leopard seals are themselves dependent on krill for food. Who do you think would have more fluorine built up in its tissues: a young leopard seal pup or an adult seal? Justify your answer with three reasons.

Name Period Date

CHAPTER 16 | INVASION OF THE AFRICAN HONEY BEE

Pre-AP Activity

In Chapter 16 you have learned how introduced species (also known as "alien" and "non-native" species) such as kudzu and the Burmese python have had dramatic impacts on their new ecosystems. Such species are usually referred to as invasive species. While the words *invasive* and *invasion* suggest a purposeful, aggressive movement into a new territory, often the species themselves are transported from their native habitat by humans. Sometimes this happens accidentally, as with the brown tree snake that has wreaked havoc on the ecology of Guam. And sometimes this introduction is very deliberate.

THE AFRICANIZATION OF THE EUROPEAN HONEY BEE

In Brazil in 1956, a prize-winning geneticist named Warwick Kerr was sent to Africa to collect queen East African honey bees. The idea was to bring the queens back and interbreed them with the European honey bees which had been introduced in South America years earlier. The European species' production of honey had been disappointing, possibly due to the tropical climate. Kerr and others thought that the African species might be better suited to Brazil. Kerr delivered 63 live queen bees to Brazil, 48 of which survived into 1957 and were mated with European honey bee drones. The "Africanized" hybrid offspring, including a number of queens, were placed in hives fitted with devices that prevented queens from escaping. Then one day in October of 1957, a beekeeper who didn't know anything about the experiment happened to see the devices. He removed them, and 26 Africanized honey bee queens escaped with swarms of drones and worker bees into the forest. By the early 1960s there were reports of swarms of honey bees attacking livestock, pets, and humans, sometimes with fatal results. These bees were markedly more defensive than their European cousins. Biologists realized that the Africanized hybrid was spreading and successfully interbreeding with the European honey bee (EHB). By the 1980s the Africanized honey bee (AHB) had reached Mexico. In 1991, Jesus Diaz, a resident of Brownsville, Texas, became the first person to be attacked by a swarm of AHBs. Diaz survived, but other people have died as a result of their encounters, and the media took to calling the species "killer bees."

In addition to being more defensive than the EHB, the Africanized species is also outcompeting the EHB for their shared niche. In the past two decades, since the first AHBs showed up in the U.S., scientists have determined that the AHB has several advantages over the EHB:

1. AHBs grow faster, meaning a population can grow and disperse more rapidly than an EHB population.
2. EHB queens are far more likely to mate with an AHB drone than an EHB drone, meaning the next generation is more likely to be Africanized. Even when given a mixture of semen that is 50% AHB and 50% EHB, EHB queens actually choose to use the AHB semen for reproduction as much as 9 out of 10 times.
3. When new queen bees hatch, one whose father was an AHB will hatch a day earlier than one whose father was European, which gives them time to kill their would-be competitors for the role of queen.
4. AHB swarms invade EHB nests and replace the queen with their own.
5. Some African traits are dominant over European traits. This means that as interbreeding continues the hybrid species becomes more like the African ancestors that were imported to Brazil.

Name Period Date

The AHB also was helped out by another invasive species. In 1987, an Asian mite that is a parasitic feeder on honeybees was found in the American southwest, just a few years before the arrival of the AHB. The mites essentially wiped out the feral European honey bee population, making it that much easier for the Africanized honey bee to move into the niche. Overall, since it first escaped into the wild in 1957, the AHB has been able to spread at a rate of 200-300 miles per year. The spread of the AHB may be limited by cold temperatures and steady precipitation, but in 2005 it was found in southern Florida, and scientists wonder if climate change and accidental transport (shipping containers) might make it easier for the AHB to continue its invasion of the United States.

Look at the map of Africanized honey bee distribution in the southwest and answer the questions below.

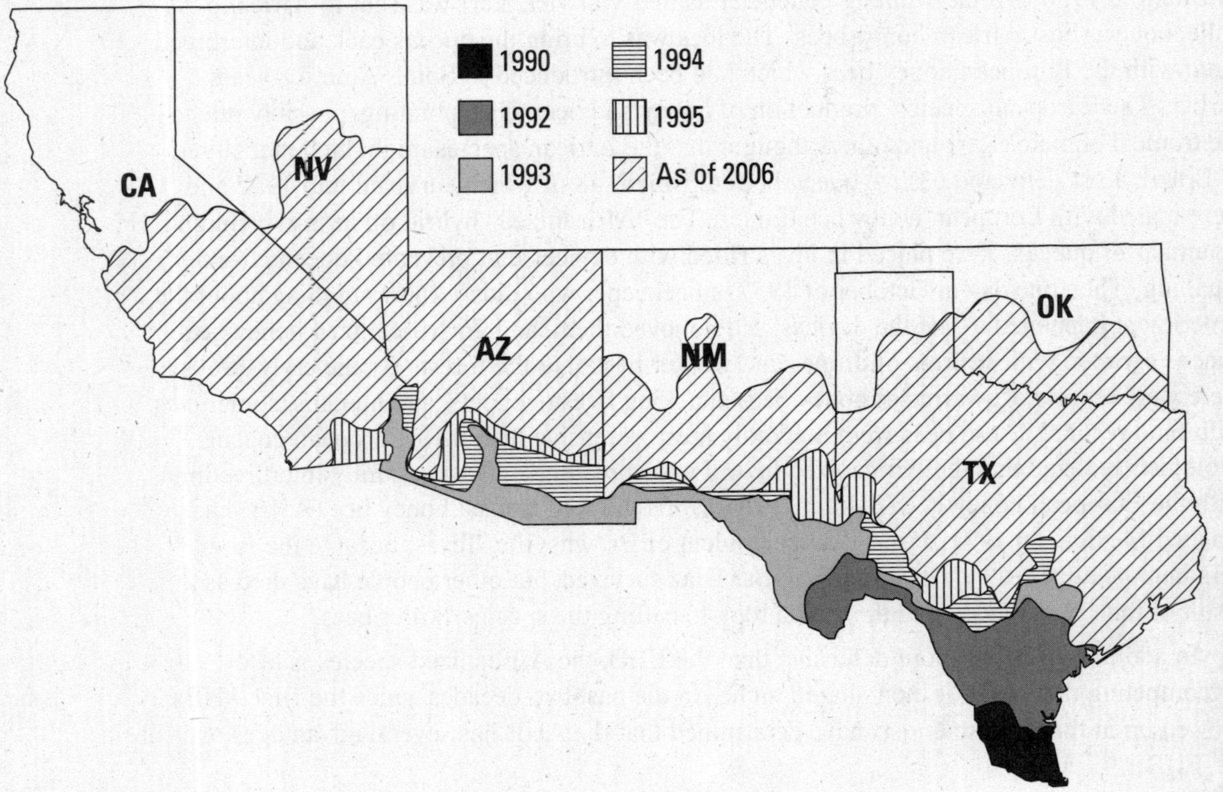

Answer the following questions on a separate piece of paper.

1. What pattern is reflected in these maps?
2. Which southwest states did the AHB colonize between 1995 and 2006?
3. Given the advantages that the AHB has over the EHB, do you think that in the future there will be many European honey bees left in the wild in North and South America? Explain.
4. How might climate change affect the range of the AHB in the United States?

CHAPTER 16
Human Impact on Ecosystems

Name Period Date

CHAPTER 16 | HUMAN IMPACT ON ECOSYSTEMS

Vocabulary Practice

nonrenewable resource	particulate	biomagnification
renewable resource	acid rain	habitat fragmentation
ecological footprint	greenhouse effect	introduced species
pollution	global warming	sustainable development
smog	indicator species	umbrella species

A. Categorize Words Write "R" next to words that can describe renewable resources. Write "N" next to words that can describe nonrenewable resources.

1. ____ wind ____ sunlight ____ oil

2. ____ coal ____ petroleum ____ water

3. ____ forest ____ deer ____ fish

4. What is the difference between a renewable and a nonrenewable resource?

____

____

B. Who Am I? Choose among these terms to answer the riddles below:

ecological footprint	indicator species	smog
global warming	introduced species	umbrella species

1. I am an organism that was brought into an ecosystem by humans and I can cause a lot of damage to native plants and animals that already live there: ____________

2. I am the amount of land required to produce and maintain enough food and water, shelter, energy, and waste to support each person on Earth: ____________

3. I am a type of air pollution: ____________

4. I am a species that is sensitive to environmental changes and can provide a sign of the quality of my ecosystem's environmental conditions: ____________

5. I am the trend of increasing global temperatures: ____________

6. I am a species that, if protected, will cause a number of other species to be protected as well: ____________

Name Period Date

VOCABULARY PRACTICE, CONTINUED

C. Matching Write the vocabulary term next to its definition.

acid rain	biomagnification	particulate	pollution

_______________ **1.** Any undesirable factor added to the air, water, or soil.

_______________ **2.** The process in which fat-soluble pollutants move from one organism to another, increasing in concentration as it moves up the food chain.

_______________ **3.** A microscopic bit of dust, metal, or unburned fuel.

_______________ **4.** A type of precipitation produced when pollutants in the water cycle cause rain pH to drop below normal levels.

ecological footprint	global warming	greenhouse effect	nonrenewable resource

_______________ **5.** Occurs when CO_2, water, and methane molecules absorb energy reradiated by Earth's surface and slow the release of this energy from Earth's atmosphere.

_______________ **6.** The amount of land necessary to produce and maintain enough food and water, shelter, energy, and waste to support each person on Earth.

_______________ **7.** The trend of increasing global temperatures.

_______________ **8.** Resources that are used faster than they can form.

habitat fragmentation	indicator species	introduced species	sustainable development

_______________ **9.** Occurs when a barrier forms that prevents an organism from accessing its entire home range.

_______________ **10.** A practice in which natural resources are used and managed in a way that meets current needs without hurting future generations.

_______________ **11.** Any organism that was brought to an ecosystem as a result of human actions.

_______________ **12.** A species that provides a sign of the quality of an ecosystem's environmental conditions.

Name Period Date

D. Vector Vocabulary Define the words in the boxes.

POLLUTION

1. ______________________

WATER POLLUTION

2. ______________________

GREENHOUSE EFFECT

3. ______________________

AIR POLLUTION

4. ______________________

ACID RAIN

5. ______________________

GLOBAL WARMING

6. ______________________

SMOG

7. ______________________

E. Crossword Puzzle Use the clues to solve the puzzle.

Across

2. Type of species that is sensitive to changes in its environment

5. Type of precipitation with a low pH caused by pollutants in the air

7. A species whose protection results in the protection of a number of other species

8. Process that results in a high concentration of pollutants in the body of a tertiary consumer

9. Kudzu in the United States

Down

1. A process that keeps heat from escaping Earth's atmosphere

3. Trend of increasing global temperatures

4. A tiny bit of dust, metal, or unburned fuel in the air

6. Smog, acid rain, or trash on a beach

10. Brown haze in the air caused by pollution

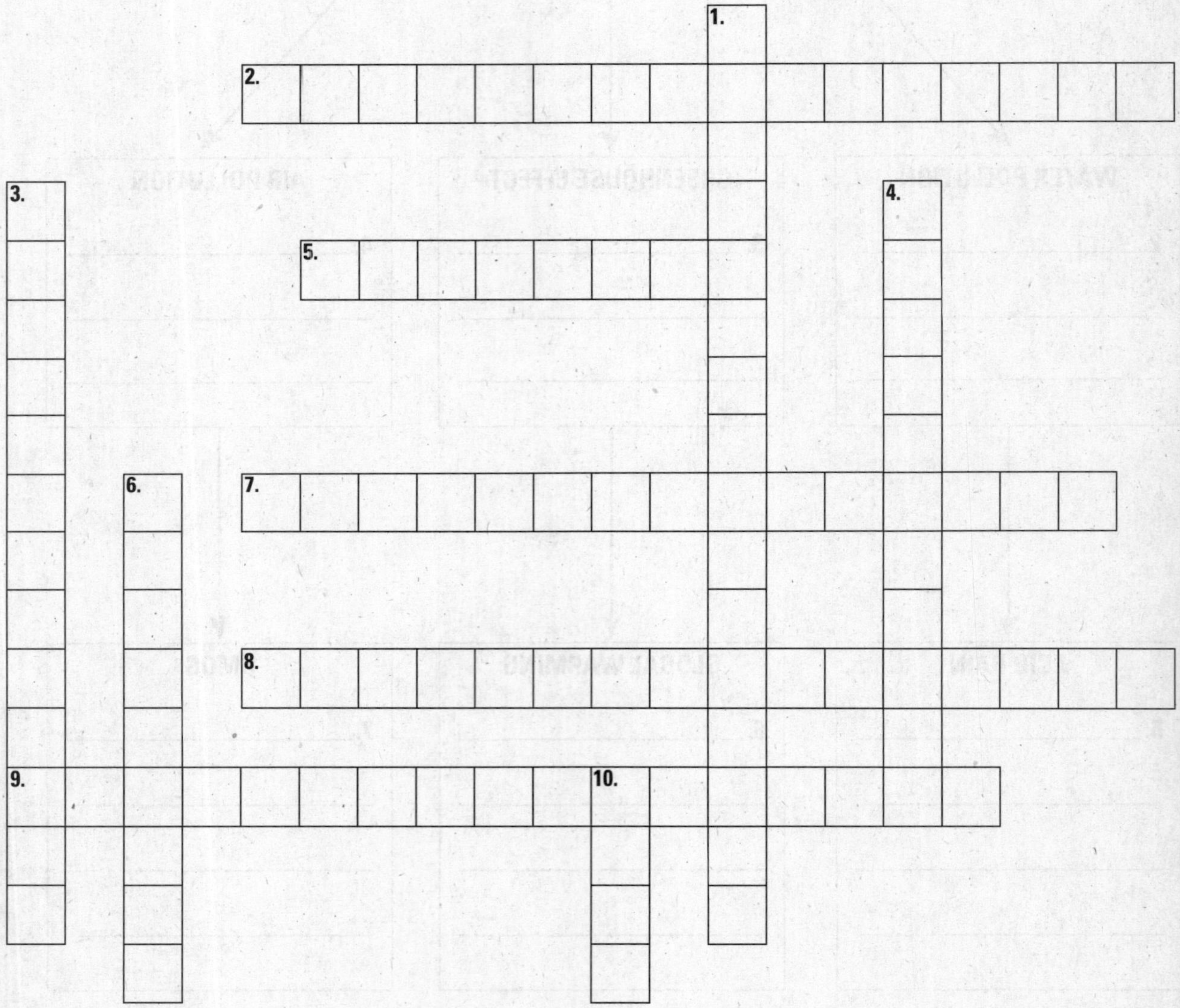

UNIT 5 | A SCHOLARSHIP FOR STEWARDSHIP

Unit Project

Suppose a national environmental magazine has offered a $1000 scholarship for a winning article on biodiversity, conservation, and environmental stewardship. The article, which will be published in their magazine, is to report on the value of biodiversity and the ecology of an endangered species in the writer's state. The article should include a conservation or recovery plan for the endangered species. You have decided to enter the competition by creating a magazine article, complete with pictures, that describes an endangered species in your state, its population status and distribution, and its habitat and niche. You will include the threats that endanger the species, a conservation or recovery plan, and how the survival of this species is beneficial to people. The endangered species you choose is not limited to animals.

- Use Internet, textbook, and/or library resources to select an interesting endangered species in your state. You must have at least three cited sources included at the end of your article. Go to **ClassZone.com** for helpful resources and links.
- Research the ecology of the species you have selected, such as its description and taxonomy, where and how it lives, and how it interacts with other species in its biological community. Also learn why it is listed as endangered and what threatens its survival.
- Read about current recovery plans for the species. Does the plan take into account all of the threats to which the species is exposed? Decide what you might do to improve the plan and how you would implement any changes to the plan. If the plan is a good one, then decide what you or your community can do to help carry out the plan. For example, if the species is being affected by water pollution, could your community help by campaigning for cleaner water? Describe and justify at least two ways you could improve or support the recovery plan and how you or your community could best carry out the measures.
- Explain how the survival of this species is beneficial to people.
- Organize your findings. Outline and fill in your notes as follows:

Title
Introduction
Description and taxonomy
Population status and distribution
Ecology, habitat, niche
Threats to survival
Conservation or recovery strategies
  Goals
  Benefits
Importance to people

- Write or type your magazine article on 8.5" x 11" unlined paper. Your article should be a minimum of four pages in length. Illustrate your article, and create an attractive title or cover page that includes your name as the writer. On the last page, list your sources of information. Bind your pages together magazine style.
- Be as creative as possible, and be sure to make your magazine article neat, organized, and easy to follow.

Name Period Date

UNIT 5 Ecology

A Scholarship for Stewardship Evaluation Rubric

Requirements	Maximum Points	Earned Points (teacher to fill in)
Endangered species is described fully, including its description, taxonomy, population status, distribution, ecology, habitat, and niche.	10	
Threats to the survival of the species are thoroughly explained, including the reason it is listed as endangered.	10	
Description of the existing recovery plan for the species is included.	15	
At least two ways you could improve or support the existing recovery plan and how you or your community could best carry out the measures are described in detail and a rationale is provided as to how this improvement or support is helpful to the species. (Two ways to improve, two ways to support, or one of each is expected.)	15	
Explanation of how the survival of this species is beneficial to people is thorough.	15	
Headings in the given outline have been incorporated into the article, and addressed.	10	
Article is illustrated and has an attractive cover that includes the magazine title, the article title, and the author's name.	10	
Magazine is bound neatly in magazine style.	5	
Article is neat, organized, and easy to follow.	5	
At least 3 cited sources are included at the end of the article.	5	
Rubric Score:	100 points	
Extra Credit (given at teacher's discretion):		
Powerpoint slideshow about the species is created and presented to the class.		
	Total Score:	
Teacher's Comments:		

UNIT
5

A SCHOLARSHIP FOR STEWARDSHIP
Unit Project Teacher Notes

Purpose: To consider an endangered species in the context of its benefit to biodiversity and give students an opportunity to analyze conservation or recovery plans for an endangered species.

Overview: Students will investigate an endangered species in their state and learn what measures are being taken to help the species survive. Students will

- search Internet, textbook, and/or library resources about an endangered species in their state
- analyze an existing recovery plan for the endangered species
- formulate a plan for improvement or support of the existing recovery plan, including how it would best be implemented
- prepare a report in an illustrated, titled, magazine-style article about their findings

Preparation:

- Copy the project description and the rubric for students.
- Plan timetable.
- You may want to locate and list several appropriate Web sites that provide the information students will require. Go to **ClassZone.com** for helpful resources and links.

Project Management:

- Assign the project at the beginning of Chapter 13.
- Have students read over the project sheet before beginning their research.
- Projects are completed outside of class and should take three weeks to complete.
- Projects should be done individually.
- Provide library references.
- All endangered species have a recovery plan.
- Have students check in weekly to monitor their progress.

Differentiation: This project can be adapted for various ability levels within the class.

- **Below Level students:** Students may complete this project by preparing a report in magazine style that describes an endangered species, its ecology, why it is endangered, and two ideas about what they might do to help the species survive. Do not include the analysis of the recovery plan.
- **Pre–AP students:** Have students include a section in the magazine detailing the Endangered Species Act and the process of getting an organism listed as an endangered species.

Answer Key

Section 13.1

Study Guide

1. **Description:** an individual living thing; **Example:** any individual organism, such as a moose
2. **Description:** a group of the same species that lives in one area; **Example:** any group of animals of the same species, such as a herd of moose
3. **Description:** group of different species that live together in one area; **Example:** any groups of different species that live in the same area, such as herds of moose and bison
4. **Description:** includes all of the organisms as well as the climate, soil, water, rocks, and other nonliving things in a given area; **Example:** any ecosystem, such as the Greater Yellowstone Ecosystem
5. **Description:** a major regional or global community of organisms; **Example:** any biome, such as a temperate grassland
6. observation is the act of carefully watching something over time
7. direct surveys are used for animals that are easy to follow and involve watching animals with the naked eye or with tools such as binoculars or scopes; indirect surveys are used for species that are difficult to track and involve searching for other signs of an animal's presence, such as looking for feces or a recent kill
8. **Laboratory:** benefit: able to control variables, drawback: experiments are performed in artificial settings that may not completely reflect a real setting; **Field:** benefit: more accurately reflects real conditions, drawback: harder to control all variables
9. scientists might want to use a model when their experiment is not practical to perform in real-time, such as when attempting to forecast how conditions might change in the future; models are also used to predict how changing one or more variables may affect future conditions
10. Ecology is the study of the interactions among living things, and between living things and their surroundings.
11. a biome contains both ecosystems and communities of organisms

Power Notes

Levels of Organization, L-R: organism, population, community, ecosystem, biome
Research methods, top to bottom, L-R: observation, experiment, modeling; direct, indirect, lab, field, mathematical, computer

Reinforcement

1. ecology is the study of the interactions among living things, and between living things and their surroundings
2. organism, population, community, ecosystem, biome
3. observation, experimentation, modeling
4. when the question he or she wishes to answer cannot be easily answered through observation or experimentation

Section 13.2

Study Guide

1. living, nonliving
2. Biotic, plants, animals
3. Abiotic, moisture, temperature, wind
4. Biodiversity is the variety of organisms found within a specific area.
5. keystone species
6. When they build dams, beavers change free-flowing stream habitats into ponds, wetlands, and meadows. This change provides habitat for a number of different species, including fish, birds, and insects, increasing the area's biodiversity.
7. Abiotic factors are nonliving things, biotic factors are living things
8. a keystone species is a species that holds together a healthy ecosystem; its

presence has a large impact on the rest of the ecosystem

Be Creative: drawings will vary; biotic factors may include plants and animals; biotic factors may include soil, Sun, precipitation

Power Notes

Ecosystem
top to bottom, L-R: biotic factor, abiotic factor; plant, animal, wind, sunlight, soil
Biodiversity is the assortment, or variety, of living things in an ecosystem
A keystone species is a species that has an unusually large effect on its ecosystem

Reinforcement

1. a living part of an ecosystem such as a tree or frog
2. a nonliving part of an ecosystem such as wind or soil
3. biodiversity is the assortment of living things in an ecosystem
4. areas of high biodiversity have a large variety of species that live near one another; preserving these areas preserves a large number of species

Section 13.3

Study Guide

1. Producers, nonliving, autotrophs
2. Consumers, eating, heterotrophs
3. Producers provide the basis for an ecosystem's energy.
4. Most producers need sunlight to make food, and consumers are dependent on producers to provide the base of the food chain in an ecosystem. Consumers are therefore indirectly dependent on the sun for their energy as well.
5. **Photosynthesis:** process in which carbohydrates are formed, energy is obtained from sunlight;**Chemosynthesis:** process in which carbohydrates are formed, energy is obtained from chemicals; **Both:** process in which carbohydrates are formed
6. An autotroph makes its own food (nourishment), while a heterotroph must get nourishment from other resources
7. in photosynthesis, energy is obtained from sunlight, in chemosynthesis, energy is obtained from chemicals
8. a producer gets its energy from nonliving resources while a consumer gets its energy by eating other living or once-living organisms

Power Notes

Producer and Consumer
,by row:
producer/autotroph, get energy from nonliving resources; consumer/heterotroph, get energy by eating other organisms
Processes by Which Producers Obtain Energy
Photosynthesis: sunlight as source of energy; green plants, cyanobacteria, some protists
Chemosynthesis: chemicals as source of energy; deep-sea organisms, hydrothermal pools, marsh flats
Both: producers make their own energy

Reinforcement

1. a producer is an organism that gets its energy from nonliving resources; a consumer gets its energy by eating other living or once-living organisms
2. producers provide the basis for an ecosystem's energy
3. consumers are indirectly dependent on the Sun since some consumers rely on producers as their main food source, and most producers rely on the Sun as their energy source.
4. Photosynthesis uses the Sun as the primary source of energy, chemosynthesis uses chemicals as the primary source of energy

Section 13.4

Study Guide

1. producer, consumer, ecosystem
2. herbivore
3. carnivore
4. omnivore
5. detritivore

6. decomposer
7. primary consumer
8. secondary consumer
9. tertiary consumer
10. trophic levels
11. A food web shows the complex network of feeding relationships within an ecosystem, while a food chain is simpler, showing only a single chain of producers and consumers.
12. Some energy is stored within an organism, and some energy is dissipated into the environment.
13. producer
14. a specialist is a consumer that primarily eats one specific, or particular, organism or feeds on a very small number of organisms, while generalist is a consumer that has a general, or varying diet, and doesn't rely on any one organism on which to feed
15. herbivores eat plants, carnivores eat meat, omnivores eat both plants and meat

Power Notes

Types of Consumers
herbivore—eats only plants
carnivore—eats only other animals;
omnivore—eats both plants and animals;
detritivore—eats dead organic matter
A food web shows the complex network of feeding relationships and flow of energy

Trophic Level:
Producer: phytoplankton, algae
Primary Consumer: zooplankton, shrimp, sea turtle, parrotfish
Secondary Consumer: sponge, jellyfish, triggerfish
Tertiary Consumer: Reef shark
Arrows point from phytoplankton to sea sponge, zooplankton, shrimp; zooplankton to sea sponge, jellyfish; algae: parrotfish, sea turtle; shrimp: jellyfish, triggerfish; triggerfish: reef shark; parrotfish: reef shark

Reinforcement

1. herbivore, carnivore, omnivore, detritivore
2. a specialist will only eat a few select organisms while a generalist will eat a wide variety of organisms
3. producer–primary consumer (herbivore)–secondary consumer (carnivore that eats herbivore)–tertiary consumer (carnivore that eats carnivore)
4. a food chain is a simple model that links species by their feeding relationships, one to the next; a food web is a model that shows the complex network of feeding relationships and the flow of energy within and sometimes beyond an ecosystem
5. some energy is stored within an organism and some energy is dissipated into the environment

Section 13.5

Study Guide

1. **Description:** process in which water or snow falls to Earth
2. **Description:** process in which water droplets reenter the atmosphere
3. **Description:** process in which water vapor is released from plant leaves
4. **Description:** process in which water vapor in the atmosphere condenses to form clouds
5. oxygen, respiration
6. photosynthesis, carbon dioxide, respiration
7. carbon dioxide
8. respiration, decomposition
9. Combustion
10. nitrogen fixation is a process in which certain types of bacteria convert gaseous nitrogen into ammonia
11. *(1)* Phosphate is released by the weathering of rocks. *(2)* Plants and some fungi take up phosphate in their roots. *(3)* Phosphorus moves from producers to consumers via the food web. *(4)* During decomposition, phosphorus is returned to the soil. *(5)* Some phosphorus leaches into the water supply, and may become locked into

sediments at the bottom of the water body. Over time, the sediments form rocks, and the cycle starts again when phosphate is released by these rocks when they erode.

12. a biogeochemical cycle is the movement of a particular chemical through the biological (living), geological (nonliving/earth), and chemical parts of an ecosystem

13. the movement of water from the atmosphere to the surface of Earth, below ground, and back

Power Notes

Boxes can be filled with words or sketches describing each cycle.

Oxygen cycle: oxygen needed for cellular respiration; respiration and photosynthesis are major components of the cycle

Carbon cycle: building block of life; respiration, decomposition, combustion, photosynthesis are major components of the cycle

Hydrologic cycle: pathway of water on Earth; condensation, precipitation, storage, evaporation, and transpiration

Nitrogen cycle: nitrogen fixation converts nitrogen into ammonia, which can be used by other organisms

Phosphorus cycle: phosphate released by the weathering of rocks, plants and fungi take phosphate in roots, phosphorus moves through food chain, returned to soil or water by decomposition

Reinforcement

1. the circular pathway of water from the atmosphere, to Earth's surface, below ground, and back into the atmosphere

2. a number of chemical elements are essential to the structure and function of organisms

3. photosynthesis and respiration

4. a process during the nitrogen cycle in which certain types of bacteria convert gaseous nitrogen into ammonia

Section 13.6

Study Guide

1. biomass

2. heat, waste

3. *From bottom of pyramid to top*: producers, primary consumers, secondary consumers, tertiary consumers

4. an energy pyramid is a diagram that compares energy used by producers, primary consumers, and other trophic levels

5. a biomass pyramid is a diagram that compares the biomass of different trophic levels within an ecosystem

6. a pyramid of numbers is a diagram that shows the numbers of individual organisms at each trophic level in an ecosystem

7. biomass is a measure of the total dry mass of organisms in a given area

8. answers will vary depending on ecosystem

Power Notes

Trophic level (top to bottom): tertiary consumer, secondary consumer, primary consumer, producer

Energy pyramid: diagram that compares energy used by producers, primary consumers, and other trophic levels

Two other pyramid models: Biomass pyramid measures dry mass at each trophic level; Pyramid of numbers measures the number of individuals at each trophic level

Reinforcement

1. an energy pyramid is a diagram that shows the distribution of energy among trophic levels

2. a measure of the total dry mass of organisms in an ecosystem

3. some of the energy is incorporated into the organism, and some of the energy is lost as heat and waste

4. a biomass pyramid compares the biomass of different trophic levels within an ecosystem and a pyramid of numbers shows the number of individual organisms at each trophic level in an ecosystem

Chapter 13

Data Analysis Practice

1. 450/22 = 20.45 = N 960/1 = 960 = A T = 20.45(960) = 19, 636.36 sea sponges or 19,636 sea sponges (rounded to the nearest whole number.)
2. Her population estimate would no longer be valid. The number she calculated would most likely be an overestimation of the population, since a portion of the sea sponge habitat was destroyed by the hurricane.

Pre-AP* Activity

DESIGN A MARK-RECAPTURE STUDY

1. Students should suggest a means of tagging or marking the wildebeests (i.e., an implant or electronic transponder that could be put beneath the wildebeest's fur or skin) that does not increase or decrease the marked animal's odds of being preyed upon, counted, or recaptured. Designs should ensure that both the marked and unmarked wildebeests have the same chances of being preyed upon and recaptured throughout the duration of the study period. Studies should be done quickly to lessen the impact of predation on the data. Students could also suggest that some rate of predation by lions or other predators be applied to the data analysis, or that predation be monitored so that any marked wildebeests that are killed prior to the recapture phase are accounted for. Designs should suggest that the study be performed outside of the seasonal reproductive periods of the species. Both the initial capture of the wildebeests and the final recapture should be done in such a way that sampling is random, such as placing a number of teams at random places in the crater and having them capture a set number of animals.
2. 1:9. (450 out of 500 are unmarked. 50:450 = 1:9.)
3. The ratio is also 1:9 (2:18). This suggests that the tag was well designed, because it did not increase or decrease the chances of being killed by a lion.
4. 5000. (50:450 = 500:4500. 500 + 4500 = 5000.)
5. Immigration and emigration. The population would be considered open.

Pre-AP Activity

BIOMASS IN CORAL REEF ECOSYSTEMS

1. The size of the levels of the two pyramids should reflect the following average fish biomasses: NWHI: 1º– 0.8, 2º– 0.43, 3º–1.34; MHI: 1º– 0.34, 2º– 0.26, 3º– 0.04 The actual shapes of the levels may vary greatly among students' work.
2. *Sample Answer:* The average biomass of tertiary (3º) consumers in the NWHI is much greater than that of the MHI. It is also greater than the biomass of the lower trophic levels in its own ecosystem. The total average biomass in the NWHI is much greater than that of the MHI.
3. fishing
4. The secondary consumers may have much shorter lives than those above them, meaning several generations of fish may feed the top level. Also, some fish migrate, meaning they could be counted in an ecosystem one day only to be living and feeding in another the next day.

Vocabulary Practice

A. Synonyms or Antonyms

1. synonym
2. antonym
3. antonym
4. synonym
5. antonym
6. antonym

B. Stepped-Out Vocabulary

1. A species that has an unusually large effect on its ecosystem; its loss greatly impacts its ecosystem; a beaver is a keystone species
2. An organism that eats both plants and animals;

omnivores are often generalists; most humans are omnivores

3. An organism that breaks down organic matter into smaller compounds; a type of detritivore; fungi are decomposers

C. Word Origins

1. the study of our home (Earth)
2. photosynthesis uses light energy to join together chemical compounds to form carbohydrates
3. a carnivore eats flesh
4. an herbivore eats vegetation
5. a detritivore breaks down organic matter into smaller pieces
6. an omnivore eats all things, plants and animals
7. chemosynthesis uses chemical energy to join together chemical compounds to form carbohydrates
8. an autotroph makes its own nourishment
9. a heterotroph eats other organisms to get its nourishment
10. a biogeochemical cycle follows the path of a chemical in both the living and geological (or earth) parts of an ecosystem
11. the hydrologic cycle is the water cycle
12. mass of living things in a given area

D. Categorize Words

1. wind, A; sunlight, A; deer, B
2. soil, A; sunflower, B; water, A
3. fungus, B; snow, A; eagle, B
4. temperature, A; prairie dog, B; frog, B

E. Find the Odd Word

1. plant; a carnivore is a consumer
2. producer; a decomposer is a type of detritivore
3. autotroph; both omnivores and herbivores are consumers (heterotrophs)
4. keystone species; an energy pyramid is made up of different trophic levels

F. What's the Difference?

1. a producer makes its own food from nonliving resources; a consumer gets its energy by eating other organisms
2. chemosynthesis is the process by which an organism forms carbohydrates by using chemicals as an energy source; in photosynthesis, light is used as the energy source
3. a food chain shows a simple sequence that links one producer to one consumer and so on; a food web shows the complex network of feeding relationships within an ecosystem
4. a community is a group of different species that live together in one area; an ecosystem includes communities of different species along with all the abiotic factors within the area as well

G. Crossword Puzzle

Across

1. energy pyramid
3. trophic level
6. biogeochemical cycle
7. ecosystem
8. biodiversity
10. biome

Down

2. nitrogen fixation
4. decomposer
5. specialist
6. biomass
7. ecology

Section 14.1

Study Guide

1. a habitat is all of the biotic and abiotic factors in the area where an organism lives, while a niche includes all physical, chemical, and biological factors that a species needs to survive, stay healthy and reproduce
2. habitat: food, other lions, trees, watering hole, wildebeest, zebra, sand, temperature, grass, savanna; niche: all of the above plus hunting behavior
3. a principle that states that when two species are competing for the same resources, one species will be better suited to the niche and the other species will either be pushed into another niche or become extinct
4. one competitor is pushed out of a niche by another competitor, niche partitioning (resources

are divided among competitors), evolutionary response (divergent evolution occurs)

5. species that occupy similar niches but live in different geographical regions
6. ecological equivalents live in two different geographic locations and therefore do not compete for the same resources
7. a habitat is an area or environment where an organism or ecological community normally dwells, or lives
8. two species that use the same resources in the same way compete together; the species that is less suited (less well-adapted) will be pushed out of (excluded from) the niche or become extinct
9. in math an equivalent refers to two sets that have identical or corresponding parts; in a similar way, ecological equivalents are two species that occupy identical niches but occur in different regions

Power Notes

Habitat: all the biotic and abiotic factors in the area where an organism lives
Ecological niche: all of the physical, chemical, and biological factors that a species needs to survive
Competitive exclusion is a principle that states when two species are competing for the same resources, one species will be better adapted to the niche, and the other will be pushed into another niche or go extinct
Two other results of competitive exclusion: niche partitioning, evolutionary response—divergent evolution
An ecological equivalent is species that occupy similar niches but live in different geographical regions

Reinforcement

1. a niche is a part of a habitat; a habitat is all of the living and nonliving factors in the area where an organism lives, while a niche includes all the specific physical, chemical, and biological factors needed by an organism to survive, stay healthy, and reproduce
2. one species will go extinct; the resources of the niche will be divided and the two species will coexist; an evolutionary response will result in selection of different traits that are successful in different parts of the niche
3. while the two species occupy similar niches, they live in different geographical regions

Section 14.2

Study Guide

1. intraspecifc
2. interspecific
3. interspecific
4. intraspecific
5. intraspecific
6. drawings will vary, may include one animal chasing another, one animal eating another, among other acceptable answers
7. **Mutualism:** *A* benefits, *B* benefits; **Commensalism** *A* benefits, *B* no impact; **Parasitism:** *A* benefits, *B* harmed
8. both refer to relationships in which one organism benefits and the other is harmed/eaten; parasitism is a symbiotic relationship in which two organisms live in close contact with one another, such a close relationship is not necessary for predation
9. an endoparasite lives on the inside of its host and an ectoparasite lives on the outside of its host
10. symbiosis is an ecological relationship between members of at least two different species that live together in direct contact
11. mutual refers to a relationship, such as a mutual agreement; mutualism is a symbiotic relationship in which both organisms benefit
12. sketches will vary

Power Notes

Competition: two organisms fight for the same limited resources;
Symbiosis: close

relationship between two or more different species that live in close contact; Predation: one organism captures and eats another organism
Mutualism: both organisms benefit +/+;
Commensalism: one organism benefits, the other is unaffected +/0;
Parasitism: one organism benefits and the other is harmed +/-

Reinforcement

1. food, access to mates, territory
2. mutualism, commensalism, parasitism
3. predation is the process by which one organism captures and feeds on another organism; parasitism is like predation in that it involves feeding on another organism, but parasitism also involves two species having a very close relationship with one another, predation does not necessarily require a close relationship

Section 14.3

Study Guide

1. number of individuals/area (units^2)
2. lack of predators, plenty of food resources
3. Refer to Visual Vocab on page 437 for visual answers; clumped, uniform, random
4. clumped dispersion—individuals may live close together in groups to facilitate mating, gain protection, or access food resources; uniform dispersion—territoriality and intraspecies competition for limited resources lead to individuals living at specific distances from one another
5. a reproductive strategy is the way an animal reproduces; for an animal with many predators, it makes sense to lay thousands of eggs because that ensures that a least a few might survive to adulthood; those animals with fewer predators can invest more time in caring for their young
6. Type I
7. Type II
8. Type II
9. Type II
10. Type III
11. Type III
12. Type I
13. Type I
14. population density is a measurement of the number of individuals living in a defined space while a population dispersion is the way in which individuals of a population are spread in an area or a volume

Power Notes

Population density is a measurement of the number of individuals living in a defined space
Formula: # of individuals/area (units^2) = population density
Population dispersion is the way in which individuals of a population are spread in an area or volume
Three dispersion types: clumped, uniform, random (with proper sketch above each label)
A survivorship curve is a generalized diagram of surviving members over time from a measured set of births
Graph should look like the one in Figure 14.8.
Three types of survivorship curves:
Type I: low infant mortality, population survives to old age; large mammals, including humans
Type II: survivorship rate equal at all ages over life span; birds, small mammals, reptiles
Type III: high birth rate, high infant mortality rate; invertebrates, fish, amphibians, plants

Reinforcement

1. a measure of the number of individuals living in a defined area
2. 30/3 = 10 birds per km^2
3. clumped dispersion: individuals live close together in groups or packs; uniform dispersion: individuals are evenly spaced across an area, may indicate that individuals are territorial and compete

for limited resources by living at specific distances from one another; random dispersion: no distinct pattern within a specific area

Section 14.4

Study Guide

1. immigration
2. deaths
3. births
4. emigration
5. when resources are abundant, populations can grow at a more rapid pace; when resources are lacking, populations begin to decline
6. Exponential growth: J-shaped curve, occurs when a population size increases dramatically over a period of time
   Logistic growth: S-shaped curve, a population begins with a period of slow growth followed by a brief period of exponential growth before leveling off at a stable size
7. logistic growth
8. exponential growth; the population may outgrow available resources such as food
8. competition, predation, parasitism and disease
9. unusual weather, natural disasters, human activities
10. immigration is the movement of individuals into a population; emigration is the movement of individuals out of a population
11. a density-dependent limiting factors depends on the population density; a density-independent limiting factor does not depend on population density
12. exponential growth occurs when a population increases rapidly over a short time; logistic growth occurs when a population grows slowly for awhile, has a short period of exponential growth, then levels off at a stable size

Power Notes

Four factors that affect the size of a population: immigration, births, emigration, death

Exponential and Logistic growth graphs should be similar to those in Figure 14.9 and Figure 14.11

Density-dependent limiting factors: competition, predation, parasitism/disease

Density-independent limiting factors unusual weather, natural disasters, human activities

Reinforcement

1. immigration, births, emigration, deaths
2. carrying capacity is the maximum number of individuals of a particular species that the environment can normally and consistently support; carrying capacity is a characteristic of logistic growth
3. a density-dependent limiting factor is any environmental influence that directly affects a population size; density-independent limiting factors limit the growth of a population regardless of its density

Section 14.5

Study Guide

1. the sequence of biotic changes that regenerate a damaged community or create a community in a previously uninhabited area
2. *(1)* 0–15 years, moss, lichens, and other pioneer species grow; *(2)* 15–80 years, shrubs, cottonwoods, and alder thickets grow; *(3)* 80–115 years, transition to forest; *(4)* 115–200 years, hemlock-spruce forest
3. *(1)* 0–2 years, weeds and other plants grow; *(2)* 2–18 years, grass, shrubs, and pine seedlings grow; *(3)* 18–70 years, pine forest and young hardwood seedlings grow; *(4)* 70–100 years, oak—hickory forest
4. primary succession is the establishment and development of an ecosystem in an area that was previously uninhabited, the process of primary succession might be started

by glacial retreat, volcanic eruptions, or landslides. Secondary succession is the reestablishment of a damaged ecosystem in an area where soil was left intact;, the process of secondary succession might be started after a flood or fire.

5. a pioneer is the first person to settle a new area; likewise, a pioneer species is the first species to inhabit a once empty area

Power Notes

Primary succession is the establishment and development of an ecosystem in an area that was previously uninhabited

Sketches should be similar to Figure 14.16; labels: 0-15 years, 15-80 years, 80-115 years, 115-200 years

Secondary succession is the reestablishment of a damaged ecosystem where the soil was left intact

Sketches should be similar to Figure 14.17; labels: 0-2 years, 2-18 years, 18-70 years, 70-100 years

Reinforcement

1. the sequence of biotic changes that regenerate a damaged community or create a community in a previously uninhabited area

2. pioneer species are the first organisms that establish themselves in a previously uninhabited area; they begin the process of breaking down the rock into soil that can hold plants, paving the way for the establishment of other species

3. communities are constantly changing over time as conditions change; for example, the continued growth of a forest changes the plant communities that live below the trees as the amount of light that reaches the ground changes

Chapter 14

Data Analysis Practice

1. As the snowshoe hare populations increase, the lynx and coyote populations increase. (Although the lynx population grows at a faster rate than the coyote population.)

2. Given the hare population bottoms out in 1993, the snowshoe hare population would then peak eight years later, in 2001. The lynx and coyote populations would grow during the same time period, but each would peak a year after the snowshoe hares, in 2002. Then all three populations would decline, with the snowshoe hare population declining at the most dramatic rate. (The extended graph drawn by the student should reflect the above statement.)

Pre-AP Activity

BROOD PARASITISM

1. The decline of songbirds could result in spelling fewer available nests for cowbirds to parasitize. This could result in a decline in cowbird populations, which could in turn help songbirds rebound.

2. Students' experimental designs should suggest monitoring the nests of host species that have the proven ability to get rid of cowbird eggs. They should check the nests of robins every day for signs that the robins or cowbirds have laid eggs in a nest and if any eggs have been destroyed or dropped from a nest. Students should determine if there is any difference in rate of destruction of cowbird eggs (destroyed eggs per laid egg) between two groups of robin nests: those in which at least one cowbird egg is laid before the robin's eggs are laid; and those in which the robin lays her own eggs before cowbirds lay theirs. Robin nests that are never parasitized by cowbirds should be excluded from both groups.

3. It could suggest that robins are more likely to recognize a cowbird egg if there are other eggs, such as its own, that it can compare them to. Perhaps the contrast in color, size, odor, or some other stimulus tells the robin that some of the eggs in its nest are not her own.

Pre-AP Activity

THE EFFECTS OF POPULATION DENSITY AND CLIMATE ON PLAGUE

Students' diagrams should make the connection (first branch) that warm spring temperatures and increased rainfall increases ecosystem productivity, thereby providing a "sudden" increased food supply, leading to the subsequent explosive growth in rodent population size. At the same time (this should be a separate branch) the spring moisture and warm temperatures could improve conditions for fleas, leading to increased reproductive success. Thus both the rodent population and the flea population grow in size, and these events increase the likelihood of plague transmission to humans.

Vocabulary Practice

A. What's the Difference?

1. primary succession occurs in a previously uninhabited area; secondary succession occurs in an ecosystem that has been disturbed, but still has soil intact
2. a habitat includes all the biotic and abiotic factors in an area where the organism lives; an ecological niche refers to the specific chemical, physical, and biological factors that an individual needs to survive
3. logistic growth has a short period of rapid growth followed by the stabilization of population growth when the carrying capacity is reached; exponential growth refers to dramatic growth over a short period of time, and population does not stabilize but continues to grow
4. density-dependent limiting factors are affected by the number of individuals in an area; density-independent limiting factors are not affected by the number of individuals
5. mutualism is a symbiotic relationship in which both organisms benefit; parasitism is a symbiotic relationship in which one organism benefits while the other organism is harmed

B. Matching

1. symbiosis
2. parasitism
3. predation
4. mutualism
5. competition
6. commensalism
7. emigration
8. carrying capacity
9. immigration
10. population crash
11. limiting factor

C. Vector Vocabulary

1. close ecological relationship between two or more organisms that live in close contact
2. is a type of
3. is a type of
4. is a type of
5. both organisms benefit
6. one organism benefits, the other neither benefits nor is harmed
7. one organism benefits and the other is harmed
8. answers will vary; lesser long-nosed bat/saguaro cactus; human and pet dog
9. answers will vary, human and eyelash mite, clownfish and sea anemone
10. answers will vary, caterpillar and wasp, human and tapeworm

D. Secret Message

1. habitat
2. limiting
3. predation
4. pioneer
5. logistic
6. competition
7. symbiosis
8. parasitism
9. immigration
10. mutualism
11. secondary

Aldo Leopold

Section 15.1

Study Guide

1. the part of Earth where life exists; formed by all of Earth's ecosystems
2. all of Earth's water, ice, and water vapor
3. the air blanketing Earth's surface
4. the features of Earth's surface, including continents, rocks, and sea floor, and everything below the surface of Earth

5. the biota is the collection of living things that live in the in the biosphere
6. Within the biosphere, a plant grows in the ground (geosphere), during photosynthesis, oxygen is expelled into the atmosphere, precipitation (hydrosphere) is needed for the plant to grow.
7. large outer circles should be labeled "atmosphere," "hydrosphere," and "geosphere," large center circle should be labeled "biosphere," and small inset circle in center should be labeled "biota"
8. The Gaia hypothesis explains how biotic and abiotic factors interact in the biosphere. In this hypothesis, the Earth is considered to be a sort of living organism, in which the atmosphere, geosphere, and hydrosphere function together to yield a biosphere that can sustain life.
9. air
10. life
11. earth
12. water

Power Notes

Earth systems

Biosphere: part of Earth where life exists; all of Earth's ecosystems

Hydrosphere: all of Earth's water, ice, and water vapor

Atmosphere: the air blanketing Earth's surface

Geosphere: continents, rocks, sea floor on Earth's surface, and everything below Earth's surface

Scientists who contributed to the Gaia Hypothesis: James Lovelock, Lynn Margulis

Gaia Hypothesis: Earth is a kind of living organism, in which the atmosphere, hydrosphere, and geosphere act together to yield a biosphere that can support life

Reinforcement

1. biosphere, hydrosphere, atmosphere, geosphere
2. the biota is the collection of living things that lives within the biosphere
3. hypothesis that considers Earth a kind of living organism, in which the atmosphere, geosphere, and hydrosphere interact together to yield a biosphere that sustains life

Section 15.2

Study Guide

1. an area's weather may change on a daily basis, while an area's climate is the long-term pattern of weather conditions in an area
2. temperature, sunlight, water, and wind
3. Earth is heated unevenly due to its curved shape
4. seasonal change results from Earth's tilt on its axis; as Earth orbits around the Sun, different regions of the planet receive higher or lower amounts of sunlight.
5. **Location:** far northern and far southern reaches of the planet; **Characteristics:** typically cold and often below freezing
6. **Location:** surrounds the equator, running from the Tropic of Cancer to the Tropic of Capricorn; **Characteristics:** warm, moist conditions
7. **Location:** located in the broad area between the polar and tropical climate zones; **Characteristics:** summer and winter seasons of equal length
8. heating causes movement in both water and air; for example, movement of air is one factor that leads to ocean currents
9. water evaporates from open bodies of water faster than it does from soil or through plant transpiration; therefore, coastal sites in general have higher humidity levels and receive more precipitation than inland areas
10. As warm, moist air nears a mountain, it rises and cools. This cooling of air results in precipitation on the side of the mountain range facing the wind. On the downwind side of the mountains, drier air produces a rain shadow.
11. an area of decreased precipitation

12. a microclimate is the climate of a small specific place within a larger area; climate describes the prevailing weather patterns of a much larger area

13. answers will vary depending on location

Power Notes

Climate is the long-term pattern of weather conditions in a region

Microclimate is the climate of a small specific place within a larger area

Climate Zones (top to bottom): polar, temperate, tropical, temperate, polar

Factors that influence climate: sunlight, air and water movement, landmasses

Reinforcement

1. climate refers to an area's long-term pattern of weather conditions, while an area's weather is shorter-term, as it changes on a daily, or even shorter time basis

2. Earth's curved shaped causes it to be heated unevenly by the Sun, leading to different climate zones

3. polar, tropical, temperate

Section 15.3

Study Guide

1. tropical rainforests are characterized by warm temperatures and abundant rainfall throughout the year; lush thick forests; rich in biodiversity

2. tropical grasslands are warm throughout the year, with specific dry and rainy seasons, tall grasses, scattered trees and shrubs, hoofed animals dominate; temperate grassland is dry and warm during the summer, snow during the winter; short or tall grasses dependent on precipitation; many animals live underground

3. very dry climate; plants store water or have deep underground root systems; animals are nocturnal or limit their activities during the day

4. a temperate deciduous forest is hot in the summer and cold in the winter, broadleaf forests; temperate rain forest has one long wet season and a relatively dry summer, evergreen conifers dominate

5. taiga has long, cold winters and short, warm and humid summers; coniferous trees dominate

6. tundra has subzero temperatures during the long winter, little precipitation; ground is permanently frozen, only mosses and low-lying plants survive; low animal diversity

7. tropical grasslands are found in tropical climates, where the climate is warm throughout the year, with a definite dry and rainy season; temperate grasslands are found in temperate climates where the climate is dry and warm during the summer, and most precipitation falls as snow during the winter

8. hot, semi-arid, coastal, and cold

9. temperate deciduous forest receives only about 75–150 cm of precipitation over the year, while a temperate rain forest receives over 250 cm of precipitation each year

10. the ground is permanently frozen in the tundra, so plants that depend on long roots systems cannot survive

11. chaparral is characterized by hot, dry summers and cool, moist winters; the dominant plant life is small-leaved evergreen shrubs

12. neither polar ice caps nor mountains have a specific plant community, one of the defining characteristics of a biome

13. polar ice caps occur around the poles at the top and bottom of Earth

14. a mountain life zone is a specific community of species that inhabit a specific elevation on a mountain

15. deciduous tree

16. coniferous tree

17. canopy

Power Notes

Tropical rain forest: warm and rainy; biodiversity high

Tropical grassland: dry and rainy seasons; tall grasses, scattered trees and shrubs
Temperate grassland: dry and warm summer, snow in winter; short or tall grasses
Desert: very dry, arid climate; cacti have deep root systems or other adaptations
Temperate deciduous forest: hot in summer and cold in winter; broadleaf forest
Temperate rain forest: one long wet season, relatively dry summer; evergreen conifers
Taiga: long, cold winters, and short, warm, humid summers; coniferous trees
Tundra: subzero temperatures during long winter, little precipitation; permafrost, only mosses and other low-lying plants

Reinforcement

1. a major community of organisms characterized by climate conditions and plant communities that live there
2. tropical rain forest, tropical grassland, temperate grassland, desert, temperate deciduous forest, temperate rain forest, taiga, tundra
3. ice caps lack a specific plant community and mountains have changing climatic conditions as elevation increases

Section 15.4

Study Guide

1. **Depth:** constantly changing **Description:** located between the high and low tide lines; organisms must be able to tolerate a wide range of conditions
2. **Depth:** a few cm to more than 200 meters deep **Description:** extends from the intertidal zone to the edge of the continental shelf; most ocean life lives in this zone
3. **Depth:** 200–2000 meters **Description:** extends from the edge of the neritic zone to the base of the continental shelf. Water is murky; many burrowing animals live in this zone
4. **Depth:** below 2000 meters **Description:** complete darkness, deep-sea vent communities support a large number of organisms
5. the neritic zone has 40 times more biomass than the rest of the ocean; plankton make up the bulk of this biomass
6. Marine phytoplankton carry out the bulk of photosynthesis on Earth, and therefore provide most of the oxygen.
7. **Coral reef:** shallow coastal water, tropical water, high productivity; **Kelp forest:** shallow coastal water, cold water, high productivity; **Both:** shallow coastal water, high productivity
8. a coral reef is made up of coral skeletal material, which packs together over thousands of years into solid structures
9. a change in conditions, such as an increase in water temperature or pollution, can kill the algae that live together with the coral, starving the coral.
10. phytoplankton
11. zooplankton

Power Notes

1. intertidal zone (organisms must tolerate a range of conditions)
2. neritic zone (few cm to 200 m, most ocean life lives in this zone)
3. bathyal zone (200–2000 m, turbid, murky water)
4. abyssal zone (depths greater than 2000 m, complete darkness)

Two unique coastal habitats: coral reef, kelp forest

Reinforcement

1. intertidal, neritic, bathyal, abyssal
2. neritic zone
3. phytoplankton are photosynthetic and get their energy from the Sun, zooplankton are animals that get their energy by eating other organisms
4. kelp forests grow in cold water, coral reefs grow in tropical water

Section 15.5

Study Guide

1. a partially enclosed body of water formed where a river flows into an ocean
2. the mixture of fresh water from a river with the salt water from the ocean
3. photosynthetic organisms thrive in estuaries throughout the year, providing the base of the aquatic food web; large detritivore communities decompose the enormous amounts of dead plant and animal matter that builds up over time
4. estuaries are used as spawning grounds by a number of species; in the estuary's calm waters, aquatic species can lay eggs, where the young can hatch and grow before heading back out to the ocean
5. organisms must be able to withstand changing conditions such as an increase or decrease in salinity or moisture
6. areas surrounding an estuary no longer have a buffer between the land and the ocean; during a storm such as a hurricane, catastrophic flooding can result
7. a wetland is an area of land that is saturated by ground or surface water for at least part of the year
8. wetlands maintain a clean water supply by filtering dirty water and renewing underground stores of water
9. **Littoral zone:** location—between the high and low water marks along the shoreline; description—waters are well-lit, warm, and shallow; diverse set of organisms inhabit this zone.
   Limnetic zone: location—open water located further out from shore; description—abundance of plankton communities, which supports populations of fish.
   Benthic zone: location—lake or pond bottom; description—less sunlight reaches this zone; decomposers live in the mud and sand of the benthic zone.
10. a region of land that drains into a river, a river system, or another body of water
11. Tidal movements in an estuary are important for the input of a large amount of nutrients from the ocean and river systems that feed the estuary. They also play a role in the mixing of the ocean's salt water with the river's fresh water .

Power Notes

Estuary
Definition: partially enclosed body of water formed where river flows into an ocean
Description: mixture of salt and fresh water; highly productive ecosystem
Other facts: plankton-based food web, important for migratory birds, "nursery of the sea"
Threats: removal due to development means loss of coastal barrier; less protection from storms and floods
Lake Zones
Littoral—shoreline, located between low and high tide marks, waters are well-lit, warm, and shallow
Limnetic—open water zone, abundance of plankton supports communities of fish
Benthic—lake bottom, less sunlight reaches this zone, decomposers live here

Reinforcement

1. a partially enclosed body of water formed where a river flows into an ocean
2. the mixing of fresh water with salt water
3. an freshwater ecosystem that has standing or slowly flowing water running through it
4. littoral, limnetic, benthic

Chapter 15

Data Analysis Practice

1. Graph should show both sets of data with the months on the x-axis, rainfall in millimeters on the left side of the y-axis, and the average temperature in degrees Celsius on the

right side of the y-axis. The precipitation should be a bar graph and the temperature should be a line graph.

2. September, October, and December
3. Rainfall occurs the most during the months of January through June when temperatures are the warmest, and rainfall occurs the least during the months of July through October, when temperatures are cooler.

Pre-AP Activity

CLIMATE CHANGE CONTROVERSY

1. *Sample Answer:* By stating "some scientists believe" and "skeptics contend" it suggests the question is open to debate, and skeptics and scientists are given equal footing. Using quotations around global warming could imply that it doesn't exist. Saying that some ice core studies "seem" to indicate makes it sound as though the evidence is open to interpretation.
2. Students answers will vary, but they should point out that reporting that goes out of its way to find or express the "other side" of an issue may mislead people into thinking that the other side is as supported and rigorously tested as the position of the scientific consensus.
3. *Sample Answer:* If one position is vastly more supported or credible than another, then it should probably be given more time or attention in an article or television segment. Journalists should only report on "sides" if there are indeed sides to an issue. They should not go out of their way to find skeptics just to create the impression of "balance."

Pre-AP Activity

GLOBAL WARMING AND METHANE EMISSION

1. Students' loops should contain the following factors: increased greenhouse effect; increased air temperatures; increased thaw area absorbs more heat; thaw area increases; tundra methane emissions increase; atmospheric methane concentration rises. Students should use + signs to indicate positive effect of each factor on the next.

Vocabulary Practice

A. Word Origins

1. vapor (air) portion of Earth
2. water portion of Earth
3. geologic (earth) portion of Earth
4. life portion of Earth
5. bottom zone of a lake
6. lake zone (open water)
7. deep water zone of ocean
8. bottomless zone, deepest part of ocean
9. shore zone of lake
10. type of tree in which the leaves fall off
11. small surface of the Earth, climate of small area

B. Who Am I?

1. deciduous
2. taiga
3. hydrosphere
4. chaparral
5. grassland
6. kelp forest
7. estuary
8. geosphere
9. coniferous
10. tundra
11. canopy
12. atmosphere
13. watershed
14. desert
15. coral reef

C. Put It in a Box

1. *Sample answers follow for all boxes:* leaves fall off, temperate biome, autumn
2. very cold, north of taiga, few plants, mosses and lichens
3. hot, humid in summer, cool and moist in winter, shrubs
4. needles, firs, does not drop needles, cones
5. cold, south of tundra, coniferous trees, boreal forest
6. where ocean and river meet; mixing of fresh water and salt water; "nursery of the sea"
7. dry, arid climate, cacti, hot, cold, semi-arid, and coastal
8. where life exists, Earth system, contains biota
9. long term weather patterns, temperature, precipitation, relative humidity

D. Categorize Words

Ocean Zones: intertidal zone, neritic zone, bathyal zone, abyssal zone
Earth Systems: biosphere, geosphere, hydrosphere, atmosphere
Lake Zones: limnetic zone, littoral zone, benthic zone
Coastal Habitats: kelp forest, coral reef
Types of Plankton: zooplankton, phytoplankton
Biomes: desert, taiga, tundra

E. Find the Odd Word

1. tundra; coniferous trees are found in the taiga (answer could also be coniferous, both tundra and taiga are biomes)
2. desert; kelp forest and coral reefs are both unique coastal habitats

Section 16.1

Study Guide

1. over 6 billion
2. agricultural advancements such as the use of gas-powered farm equipment; medical advances such as the development of vaccines, antibiotics, and medical surgery procedures
3. renewable
4. nonrenewable
5. renewable
6. renewable
7. renewable
8. renewable
9. renewable
10. nonrenewable
11. tree; cutting down trees at a slower rate or finding an alternate resource, using seeds to plant new trees
12. the amount of land necessary to produce and maintain enough food and water, shelter, energy, and waste to support each person on Earth
13. amount of resource use, efficiency of resource use, amount of waste produced, and toxicity of waste produced
14. renewable resources are resources that cannot be used up or can replenish themselves over time; nonrenewable resources are resources that can be used up and are used at a faster rate than they form

Be Creative: poster designs will vary

Power Notes

Graph should be similar to that shown in Figure 16.1.
Two technological advancements that have contributed to population growth: gas-powered farm equipment, antibiotics and antiseptics (medical advancements)
Types of resources
nonrenewable: used faster than it forms; oil, coal
renewable: cannot be used up or can replenish itself; sunlight, wind, water
Ecological Footprint
Definition: amount of land necessary to produce and maintain enough food and water, shelter, energy, and waste
Size depends on: amount and efficiency of resource use and the amount and toxicity of waste produced

Reinforcement

1. advancements in technology such as medical advancements and the development of the gas-powered engine
2. a renewable resource can replenish itself over a short period of time while a nonrenewable resource cannot replenish itself faster than it can be formed
3. everything in a person's life that comes from a natural product

Section 16.2

Study Guide

1. any undesirable factor, or pollutant, that is added to the air, water, or soil
2. a type of air pollution caused by the interaction of sunlight with pollutants produced by fossil fuel emissions
3. particulates, or microscopic bits of dust, metal, and unburned fuel and ground-level ozone; forms when sunlight interacts with pollutants from fossil fuels
4. a type of precipitation produced when pollutants in the water cycle cause rain

pH to drop below normal levels

5. acid rain threatens water supplies and habitat; it can cause a decline in growth rates; it also affects trees by causing leaves and bark to break down more quickly, causing trees to be more vulnerable to disease and weather
6. wavelengths of the Sun's energy
7. it makes Earth suitable for life
8. carbon dioxide
9. a process that occurs when carbon dioxide, water, and methane molecules absorb energy reradiated by Earth's surface and slow the release of this energy from Earth's atmosphere
10. global warming refers to the trend of increasing global temperatures' changes in global temperature are the result of increased levels of greenhouse gases such as carbon dioxide, water, and methane that cause the greenhouse effect
11. A greenhouse is a glass house used to grow plants. The glass allows light to pass through to provide energy for plant growth, but prevents infrared radiation from escaping, keeping the inside of the greenhouse warm. This same effect occurs within Earth's atmosphere as certain greenhouse gases absorb energy and slow the release of this energy from Earth's atmosphere
12. particulates are made of small bits of dust, metal, and unburned fuel

Power Notes

Greenhouse effect occurs when carbon dioxide, water, and methane molecules absorb energy reradiated by Earth's surface and slow the release of energy from Earth's atmosphere

1. light emitted from the Sun penetrates Earth's atmosphere
2. energy is absorbed by Earth and reradiated as heat (infrared radiation)
3. greenhouse gases in atmosphere absorb longer wavelengths of infrared radiation
4. molecules re-release infrared radiation, which is reabsorbed or lost to outer space

Reinforcement

1. smog and acid rain
2. the greenhouse effect is a natural process in which greenhouse gases, such as water and carbon dioxide absorb some of the energy released by Earth's surface, which helps to keep the surface of Earth warm
3. global warming is the trend in increasing global temperatures as a result of increased levels of greenhouse gases; the same gases that retain heat in the greenhouse effect

Section 16.3

Study Guide

1. agricultural run-off from farms, raw sewage, chemical contaminants from industrial sites
2. an indicator species provides a sign, or indication, of the quality of an ecosystem's environmental conditions
3. a process in which a pollutant moves up the food chain as predators eat prey, accumulating in higher concentrations in the bodies of predators
4. See Figure 16.11 for sample visual answer; concentration of pollutants increases as you move up the food chain from herbivores to top predators
5. *bio-* means "life," *magnification* means "to enlarge the size of something"; biomagnification refers to the increase in concentration of toxins as you move up the food chain; what occurs in small concentrations at the bottom of the food chain magnifies into a much larger concentration at the top of the food chain
6. poster designs will vary

Power Notes

An indicator species is a species that provides a sign of the quality of the

ecosystem's environmental conditions
Biomagnification is a process in which a pollutant moves up the food chain as predators eat prey, accumulating in higher concentrations
Trophic level (top to bottom): quaternary consumer, tertiary consumer, secondary consumer, primary consumer, producer

Reinforcement

1. an indicator species is a species that provides a sign, or indication, that there may be a problem with pollution in an ecosystem
2. biomagnification is the process by which pollutants accumulate in larger amounts as it moves through the food chain; for example, as a food chain moves from plankton to fish to larger fish to an eagle, the concentration of pollutants increases, so that the eagle has the highest concentration of pollutants in its body
3. pollutants are more highly concentrated the higher up the food chain

Section 16.4

Study Guide

1. the wide array and assortment of species that are found in any ecosystem
2. a loss of biodiversity can reduce an ecosystem's stability and make it more difficult for the ecosystem to handle future change
3. tropical rain forests; warm temperatures and plenty of precipitation all year long provides for a long growing season that can support a large number of different species
4. urban sprawl, development of roadways, forest harvesting
5. any organism that was brought to an ecosystem as a result of human actions
6. **Burmese Python:** feeds on small animals including endangered species; **Kudzu:** deprives other plants of sunshine they need to survive; **Mice:** devour agricultural crops
7. habitat fragmentation refers to the separation of a species' habitat into much smaller places, where obstructions such as roads or other human development creates a barrier that prevents an organism from accessing its entire home range
8. Plogans will vary.

Power Notes

Why biodiversity is important:
decrease in an ecosystem's biodiversity affects all species; medical and technological advancements come from nature; loss of biodiversity can reduce an ecosystem's stability and make it more difficult for the ecosystem to handle future change
Threats to biodiversity
Habitat fragmentation: occurs when a barrier forms that prevents an organism from accessing its entire home range
Introduced species: any organism that was brought to an ecosystem as the result of human actions
Introduced species
Burmese python, Florida Everglades, feeds on native species, including endangered birds
Kudzu, southeastern United States, blankets trees and shrubs, killing native plants
Mice, Australia, feed on crops, causing millions of dollars of damage

Reinforcement

1. biodiversity is the assortment, or variety, of living things in an ecosystem
2. habitat fragmentation forms a barrier within the home range of a species, effectively cutting off an individual's access to its entire home range
3. an introduced species is any species that was brought to an ecosystem as the result of human actions

Section 16.5

Study Guide

1. sustainable development helps to ensure that Earth can continue to support, or

sustain, a growing human population by managing resources in a way that meets current needs without harming future generations

2. **Resource:** timber; **Managed:** selective cutting; **Benefits:** encourages rapid regrowth and minimally impacts forest ecosystems; **Resource:** global fisheries; **Managed:** harvest rotation, harvest reduction, fishing bans, use of less invasive fishing gear; **Benefits:** allows fish populations to rebound
3. a species for which protection means a wide range of other species will also be protected
4. umbrella species
5. the Endangered Species Act
6. protect many other species in its ecosystem
7. Clean Air Act, Clean Water Act, Endangered Species Act
8. control birth rates, develop technology to produce more food and less waste, protect and maintain ecosystems by reducing the impact of land development
9. the goal of sustainable development is to maintain the quality of ecosystems and quantity of resources in a way that provides enough for today's population while providing for future generations as well

Power Notes

Sustainable development is a practice in which resources are used and managed in a way that meets current needs without hurting future generations

Sustainable practices in the fishing industry: rotation, fishing gear review, harvest reduction, fishing bans

An umbrella species is a species whose protection means a wide range of other species will also be protected

Three important environmental laws: Clean Air Act, Clean Water Act, Endangered Species Act

Ways in which humans can protect environment: control population growth; develop technology to produce more food and produce less waste; change practices and maintain ecosystems

Reinforcement

1. sustainable development is a practice in which natural resources are used and managed in a way that meets current needs without hurting future generations
2. by protecting an umbrella species, a wide range of other species that live in the same habitat are also protected, which in turn benefits an entire ecosystem
3. Endangered Species Act, Clean Air Act, Clean Water Act

Chapter 16

Data Analysis Practice

1. The independent variable, air quality, is discrete because it is expressed in categories that cannot be broken down into smaller units. The dependent variable, days, is continuous because it is expressed in a unit that can be broken down into smaller units of time.
2. No, the dependent variable is still continuous.

Pre-AP Activity

BIOMAGNIFICATION OF FLUORINE IN PENGUINS

1. 3,000 times.
2. 0.08 kg
3. The internal temperature (38–40º C) of the penguin denatures the enzyme that releases fluorine from the cuticle (shell). Penguins lack the gastric ceca that would enable the digestion of chitin, so the shell passes through their gut. It takes only 3 or 4 hours for krill to pass through the penguin's gut, meaning there isn't a lot of time for fluorine to be absorbed or for fluorine to migrate out of the krill's shell into the flesh.
4. Removal of the shells.
5. Whales have the warm internal body temperature that denatures the enzyme

that releases fluorine from the chitinous shells from krill.

6. The blue whale is so large that it needs to eat tons of krill. This means that it is ingesting much more fluorine than a penguin ingests.
7. An adult seal. For one thing, an adult has been accumulating fluorine for years, whereas the young seal has only just begun. Secondly, the adult seal is feeding on animals high up in the food chain, some of which will have accumulated fluorine themselves. Thirdly, the adult seal is larger than the pup, so it must eat more prey (more fluorine). Students may offer other examples.

Pre-AP Activity

INVASION OF THE AFRICAN HONEY BEE

1. The movement of the AHB northward from Mexico.
2. Nevada and Oklahoma
3. Probably not. The AHB is dominant over the EHB genetically, behaviorally, and reproductively.
4. Students should suggest that higher temperatures and other climate changes could allow the AHB to spread into other areas of the U.S.

Vocabulary Practice

A. Categorize Words

1. wind, R; sunlight, R; oil, N
2. coal, N; petroleum, N; water, R
3. forest, R; deer, R; fish; R
4. a renewable resource cannot be used up or can replenish itself over time while a nonrenewable resource is used faster than it can form

B. Who Am I?

1. introduced species
2. ecological footprint
3. smog
4. indicator species
5. global warming
6. umbrella species

C. Matching

1. pollution
2. biomagnification
3. particulate
4. acid rain
5. greenhouse effect
6. ecological footprint
7. global warming
8. nonrenewable resources
9. habitat fragmentation
10. sustainable development
11. introduced species
12. indicator species

D. Vector Vocabulary

1. undesirable factor added to soil, water, or air
2. pollutant in the water
3. reabsorption of Earth's energy by greenhouse gases
4. pollutants in the air
5. type of precipitation produced when water pollutants in water cycle cause pH in rain to be lower than normal
6. increase in global temperatures, enhanced by production of greenhouse gases
7. type of air pollution caused by interaction of sunlight and fossil fuel emissions

E. Crossword Puzzle

Across

2. indicator species
5. acid rain
7. umbrella species
8. biomagnification
9. introduced species

Down

1. greenhouse effect
3. global warming
4. particulate
6. pollution
10. smog